FOOD AND WAR IN TWENTIETH CENTURY EUROPE

Food and War in Twentieth Century Europe

Edited by

INA ZWEINIGER-BARGIELOWSKA
University of Illinois, Chicago, USA

RACHEL DUFFETT
University of Essex, UK

ALAIN DROUARD
*University of Paris-Sorbonne, France
and President of ICREFH*

Routledge
Taylor & Francis Group

LONDON AND NEW YORK

First published 2011 by Ashgate Publishing

2 Park Square, Milton Park, Abingdon, Oxon OX14 4RN
711 Third Avenue, New York, NY 10017, USA

Routledge is an imprint of the Taylor & Francis Group, an informa business

First issued in paperback 2016

British Library Cataloguing in Publication Data
Food and war in twentieth century Europe.
   1. World War, 1914–1918 – Food supply – Europe. 2. World War, 1939–1945
   – Food supply – Europe. 3. Operational rations (Military supplies) – Europe –
   History – 20th century. 4. Food supply – Government policy – Europe – History –
   20th century. 5. Food habits – Europe – History – 20th century.
   I. Zweiniger-Bargielowska, Ina. II. Duffett, Rachel. III. Drouard, Alain.
   363.8'094'0904–dc22

Library of Congress Cataloging-in-Publication Data
Zweiniger-Bargielowska, Ina.
   Food and war in twentieth century Europe / by Ina Zweiniger-Bargielowska, Rachel
   Duffett, and Alain Drouard.
       p. cm.
   Includes bibliographical references and index.
   ISBN 978–1–4094–1770–5 (hbk. : alk. paper) – ISBN 978–1–4094–1771–2 (ebook)
   1. Food supply – Europe – History – 20th century. 2. Food supply – Social aspects –
   Europe – History – 20th century. 3. War and society – Europe – History – 20th century.
   4. World War, 1914–1918 – Europe. 5. World War, 1939–1945 – Europe. 6. War and
   society – Europe – History – 20th century. 7. Europe – History, Military – 20th century.
   8. Europe – Social conditions – 20th century. I. Duffett, Rachel. II. Drouard, Alain.
   III. Title.
   HD9015.A2Z88 2011
   363.8094–dc22   2011013246

ISBN 13: 978-1-4094-1770-5 (hbk)
ISBN 13: 978-1-138-26103-7 (pbk)

Contents

List of Figures

List of Tables

List of Contributors

Peter J. Atkins, Department of Geography, Durham University, Durham DH1 3LE, UK. Email: p.j.atkins@durham.ac.uk

Isabelle von Bueltzingsloewen, Université de Lyon, Laboratoire de recherche historique, Lyon, France. Email: isavonb@club-internet.fr

Alain Drouard, National Center For Scientific Research, UMR 8596 du Cnrs, Centre Roland Mousnier, Paris, 16 rue Parrot, 75012 Paris, France. Email: adrouard01@noos.fr

Rachel Duffett, Department of History, University of Essex, Wivenhoe Park, Colchester, CO4 3SQ, United Kingdom. Email: rsduffett@hotmail.com

Martin Franc, Masaryk Institute, Archives of the Academy of Sciences of Czech Republic. Gabcikova 2362/10, 182 00 Praha 8, Czech Republic. Email: francmartin@seznam.cz

Maja Godina Golija, Institute of Slovenian Ethnology, SRC SASA, Novi trg 5, 1000 Ljubljana, Slovenia. Email: maja.godina-golija@uni-mb.si

Adel P. den Hartog, Formerly Wageningen University, Division of Human Nutrition, The Netherlands. Email: apdenhartog@planet.nl

Guðmundur Jónsson, School of Humanities, University of Iceland, Arnagardur, 101 Reykjavik, Iceland. Email: gudmjons@hi.is

Örn D. Jónsson, School of Business, University of Iceland, Oddi, 101 Reykjavik, Iceland. Email: odj@hi.is

Peter Lummel, Open-air museum Domain Dahlem, Koenigin-Luise-Str. 49, 14195, Berlin, Germany. Email: lummel@domaene-dahlem.de

Kenneth Mouré, Department of History and Classics, 2–28 Tory Building, University of Alberta, Edmonton, AB Canada T6G 2H4. Email: kenneth.moure@ualberta.ca

Svend Skafte Overgaard, Metropolitan University College Copenhagen, Department of Health and Nutrition, Pustervig 8, 1126 Kbh. K., Denmark. Email: svendo@hum.ku.dk

Alicia Guidonet Riera, Departament de Desenvolupament Humà i Acció Comunitària, University of Vic, C/Sagrada Família, 7, 08500 Vic, Spain. Email: alicia.guidonet@uvic.cat

Steven Schouten, Research Fellow, Scientific Council for Government Policy, Lange Vijverberg 4–5, 2513 AC The Hague, Netherlands. Email: steven. schouten@eui.eu

Hans-Jürgen Teuteberg, Westfälische Wilhelms-Universität Münster, Historisches Seminar, Domplatz 20–22, 48143 Münster, Germany. Email: teuteberg-uni-muenster@gmx.de

Ulrike Thoms, Institute für Geschichte der Medizin, Zentrum für Human- und Gesundheitswissenschaften, Humboldt Universität/ Freie Universität Berlin, Klingsorstr. 119, 12203 Berlin, Germany. Email: Ulrike.thoms@charite.de

Pavel Vasilyev, Department of History, Central European University, 9 Nador Street, Budapest, Hungary H-1051. Email: p.a.vasilyev@gmail.com

Ina Zweiniger-Bargielowska, Department of History, University of Illinois, Chicago, 913 University Hall, 601 South Morgan Street, Chicago, IL 60607, USA. Email: inazb@uic.edu

Preface

Alain Drouard

In 2009 the ICREFH celebrated its twentieth anniversary by publishing a new book, *The Rise of Obesity in Europe*, and organizing its eleventh symposium on 'Food and War in Europe in the nineteenth and twentieth centuries' at the University of Paris-Sorbonne. By doing so, ICREFH has remained faithful to its original programme – a biennial meeting in a European city of a symposium on the subject of food history in which the focus is on discussion and exchange between participants.

An emerging area of historiography in the history of food which could no longer be ignored is the impact on food brought about by war, especially the two world wars. Wars are not only interruptions of normal life; they deeply influence the daily life while provoking severe ruptures and breaks in food production, distribution and consumption. These can have long lasting effects on the economic structure of the food industry, government food policy as well as the individual food habits. One important topic is the question of food supplies to the troops which are of course imperative during war time. Other issues which have received little attention to date include the role of the nutritional sciences in the planning of rations. Army rations have been perceived as a possibility to get people used to better and healthier food, thereby providing a means of educating mostly young men. Military historians have also recently become concerned with the question of the army's peace time diet, both in term of quantities of rations and their sources of supply.

At the same time an increasing number of studies on civilian populations have been published in which civilians are considered as full agents of history. Wars have had major consequences for civilian food and eating, both during conflicts and long after wars have ended, including postwar restrictions on supply. The return to peace has often been synonymous with changes in the diet of a population. Wars were also synonymous with innovations and food innovations resulting from the needs of armies have subsequently spread among civilian populations. Important examples are tinned and frozen food, which were first introduced and produced on a large scale to serve the need of armies. Armies also served as a field of experimentation and a kind of test market offering a possibility to develop methods of cheap mass production. Food shortages and restrictions resulted in a search for substitutes or new products and the experience of war had important consequences for public health.

The Paris symposium explored the important question of how wars and food are related and intertwined during three days from 8 to 10 September 2009. Whether historians or sociologists, anthropologists or physicians, all participants were engaged in a rich, comparative and interdisciplinary discussion.

ICREFH Symposia

The Current State of European Food History Research (Münster, 1989).

The Origins and Development of Food Policies in Europe (London, 1991).

Food Technology, Science and Marketing (Wageningen, 1993).

Food and Material Culture (Vevey, 1995).

Order and Disorder: the Health Implication of Eating and Drinking (Aberdeen, 1997).

The Landscape of Food: Town, Countryside and Food Relationships (Tampere, 1999).

Eating and Drinking Out in Europe since the late Eighteenth Century (Alden Biesen, 2001).

The Diffusion of Food Culture: Cookery and Food Education during the Last 200 Years (Prague, 2003).

Food and the City in Europe since the Late Eighteenth Century: Urban Life, Innovation and Regulation (Berlin, 2005).

From Under-Nutrition to Obesity: Changes in Food Consumption in Twentieth-Century Europe (Oslo, 2007).

Food and War in Europe in the Nineteenth and Twentieth Centuries (Paris, 2009).

The History of the European Food Industry in the Nineteenth and the first Half of the Twentieth Century (Bologna, 2011).

ICREFH Publications

ICREFH I – Teuteberg, H.J. (ed.), *European Food History: a Research Overview*, Leicester, 1992.

ICREFH II – Burnett, J. and Oddy, D.J. (eds), *The Origins and Development of Food Policies in Europe*, London, 1994.

ICREFH III – Hartog, A.P. den (ed.), *Food Technology, Science and Marketing: European Diet in the Twentieth Century*, East Linton, 1995.

ICREFH IV – Schärer, M.R. and Fenton, A. (eds), *Food and Material Culture*, East Linton, 1998.

ICREFH V – Fenton, A. (ed.), *Order and Disorder: the Health Implications of Eating and Drinking in the Nineteenth and Twentieth Centuries*, East Linton, 2000.

ICREFH VI – Hietala, M. and Vahtikari, T. (eds), *The Landscape of Food: the Food Relationship of Town and Country in Modern Times*, Helsinki, 2003.

ICREFH VII – Jacobs, M. and Scholliers, P. (eds), *Eating Out in Europe: Picnics, Gourmet Dining and Snacks Since the Late Eighteenth Century*, Oxford, 2003.

ICREFH VIII – Oddy, D.J. and Petranova, L. (eds), *The Diffusion of Food Culture in Europe from the Late Eighteenth Century to the Present Day*, Prague, 2005.

ICREFH IX – Atkins, P.J., Lummel, P. and Oddy, D.J. (eds), *Food and the City in Europe since 1800*, Aldershot, 2007.

ICREFH X – Oddy, D.J., Atkins, P.J. and Amilien, V. (eds), *The Rise of Obesity in Europe: A Twentieth Century Food History*, Farnham, 2009.

Acknowledgments

ICREFH warmly thanks Professor Alain Drouard, organizer of the symposium and the University of Paris-Sorbonne (Paris IV) which hosted the symposium and supported publication of this book with a generous subvention. ICREFH wishes to acknowledge the kind hospitality of General Robert, Chair of the Historical Center of the Defense at the Château de Vincennes. Thanks also are due to the IEHCA (Institut européen d'histoire et des cultures alimentaires), the Ministry of Higher Education and Research and the Centre for the Study of the History of Defense for supporting the organization of the symposium.

Chapter 1

Introduction

Ina Zweiniger-Bargielowska

Why Food and War

Wars cannot be fought without food and this book explores the impact of war on food production, allocation and consumption in Europe in the twentieth century. An adequate diet is of course necessary to sustain armies, but it is equally imperative to maintain civilian health and morale in the era of total war. The analysis ranges from military provisioning and systems of food rationing to civilians' survival strategies and the role of war in stimulating innovation and modernization. These issues are approached from a comparative perspective. *Food and War in Twentieth Century Europe* brings together 18 scholars from many European countries. By incorporating this wide range of research into a single volume, this book shows how the competing claims of military and civilian provisioning were dealt with in different countries and conflicts.

Food and War examines the challenges involved in feeding huge armies as well as the difficulties of administering food controls and rationing schemes on the home front. Official distribution channels were frequently bypassed in black markets and there were important differences between social groups as well as urban and rural populations in terms of access to food supplies. Food should not be understood as a static resource and this book demonstrates how new and alternative foods were developed and utilized in times of scarcity. This included technological innovation in food processing, a shift from meat to grain in the diet and the emergence of new foods such as horsemeat. *Food and War* incorporates belligerent, occupied and neutral countries. The focus is on the impact of the two world wars in Germany, France and the United Kingdom, but this is supplemented with case studies from several other countries including Denmark and Iceland and conflicts such as the Spanish Civil War.

Europe in the Era of Total War: The Importance of Food

The literature on war in Europe during the twentieth century and particularly the two world wars is enormous, but the central role of food in modern warfare has not received the attention it deserves. There are several monographs and shorter contributions, but the topic has not generally been approached from a comparative perspective. One important exception is a collection edited by Frank Trentmann and Flemming Just, *Food and Conflict in Europe in the Age of the Two World Wars*.[1] Trentmann and Just define conflict in a broad sense as competition over food allocation rather than focusing specifically on war and they approach this topic within a wider context of food scarcity between the 1890s and the 1950s. Another is *The Taste of War: World War Two and the Battle for Food* by Lizzie Collingham, which is framed within a global perspective. This book is aimed at a general readership and it draws on an impressive range of secondary literature, but there is little primary research.[2]

There are a number of studies which explore the specific problems of urban food supplies during the First World War. This includes work on Berlin, Paris and The Hague, which provides an interesting example of the impact of war on a city in a neutral country.[3] Other studies of the home front and the politics of food include Lars Lih, *Bread and Authority in Russia, 1914–21*, Margaret Barnett, *British Food Policy during the First World War*, William Moskoff, *The Bread of Affliction: The Food Supply in the USSR During World War II* and Ina Zweiniger-Bargielowska, *Austerity in Britain: Rationing, Controls and Consumption, 1939–55*.[4] *Food and War* helps to contextualize scholarship on particular cities, countries and wars by bringing together research on both world wars and other conflicts and by incorporating belligerents, occupied territories and neutral countries whose agricultural sectors and trade patterns were profoundly disrupted by war.

The role of food in wartime can be approached from a variety of perspectives. Mancur Olson's classic economic analysis of wartime shortages explores the history of British food supplies from the Napoleonic War until the Second World War.[5] Olson draws attention to the UK's exceptional dependency on food imports, which superficially left the country extremely vulnerable in contrast with self-sufficient nations or food exporters such as Russia. In reference to the First World War, Avner Offer has argued that the UK benefited from long-established trading relationships and a sophisticated distribution system, strengthened by imperial ties and military alliances, which made it possible to maintain sufficient imports in the face of blockade and submarine attacks on merchant shipping.[6] By contrast, as Paul Vincent has demonstrated Germany suffered severely as a result of the Allied blockade during the First World War.[7] Britain's imperial ties and advantageous position in the global food trading system yet again secured vital imports in the Second World War, when the UK further benefited from Lend-Lease supplies from the USA. An effectively administered policy of food controls, rationing and the

---

1   Trentmann and Just 2006.

2   Collingham 2011.

3   Davis 2000, Bonzon 2006, de Nijs 2006, Bonzon and Davis 1997.

4   Lih 1990, Barnett 1985, Moskoff 1990, Zweiniger-Bargielowska 2000, Wall and Winter 1988, see also Schmidt 2007.

5   Olson 1963.

6   Offer 1989.

7   Vincent 1985.

expansion of communal feeding, including so-called British Restaurants, further ensured that resources, which were more plentiful than elsewhere, were distributed relatively equitably.[8] Thus, the British experience was in many ways exceptional.

By contrast, hunger and starvation were rife on the European continent during both world wars. In the Second World War, the German occupation authorities steered food supplies towards the German war machine and German civilian consumption. Food functioned as a weapon and rations in occupied Europe frequently did not take nutritional requirements into account resulting in severe undernutrition and mass starvation. French and Dutch administrators aimed to prevent deterioration in public health standards, for example by means of nutrition education. The success of this endeavour was limited as demonstrated by excess mortality among vulnerable groups in occupied France and the Dutch hunger winter of 1944–45.[9] The German military cut off the food supplies to Russian cities, a million people died in the siege of Leningrad alone, and similar numbers of Soviet prisoners of war were deliberately starved to death. Another group of 'useless eaters' were Jews, millions of whom were starved or executed.[10] Occupied Greece provides another example of wartime famine in this period.[11] By contrast, the occupation of Denmark was relatively lenient and civilian food supplies were maintained despite substantial food exports to Germany.[12] The example of Iceland, where the war resulted in unprecedented economic growth as a result of high demand for Icelandic fish and other resources, provides an important counterpoint to the dominant pattern in continental Europe.

Nutrition scientists subscribed to different theories about protein requirements around the turn of the century and the question as to what an adequate diet actually consisted of was hotly debated in the wake of the discovery of vitamins.[13] British and German nutrition scientists differed in their understanding of dietary requirements during the First World War and Mikulas Teich has argued that the former had a greater influence on policy whereas David Smith has warned that this should not be exaggerated.[14] The commitment to a high protein diet and priority attached to military provisioning in Germany contributed to defeat and revolution. By contrast, the Danish rationing system based on nutritionist Mikkel Hindhede's low protein diet helped this neutral country not only to avoid starvation despite the blockade but, arguably, contributed to improvements in public health standards. Scientists played a much larger role in the Second World War, although David Smith has stressed that 'scientists' should not be seen as a monolithic group and highlighted that the relationship between scientific findings and policy decisions in

---

8   Barnett 1985, Zweiniger-Bargielowska 2000.
9   Trienekens 2000, Futselaar 2006.
10 Collingham 2011: 211, 180–218.
11 Hionidou 2006.
12 Nissen 2006.
13 Weatherall 1995, Kamminga 2000, Vernon 2007, Treitel 2008.
14 Teich 1995, Smith 1997.

Britain were complex.[15] Nazi Germany styled itself as a modern, scientific nation and the regime invested heavily in food science and technology.

Wars are agents for change, but it is important to distinguish between short-term adaptation and longer-term transformations with regard to government policy, food production and individual habits. The imperative of feeding armies and civilians in the context of wartime shortage resulted in significant technological progress for example with regard to dehydration and freezing techniques in Nazi Germany. This technology, which was developed in collaboration between the military and major industrial firms, provided West-German industry with a technological advantage during the period of postwar reconstruction. Many other substitutes and experimental foods, ranging from dried egg to ground wood in bread flour did not acquire lasting significance. While dried egg was popular in Britain during the period of rationing in the 1940s, experiments with unusual ingredients in bread flour, which included tree bark, straw and lichen in Czech lands functioned more as a rhetorical device to ridicule allegedly mad German scientists than as a genuine addition to food resources during the First World War. By contrast, a longer-term change in France was the adoption of horsemeat, first popularized during the Franco-Prussian war, which became an important source of cheap meat until the 1960s.

Finally, it is necessary to take into account cultural preferences and culinary traditions. Horsemeat was not generally adopted throughout Europe, although horses and also dogs and cats were consumed in emergencies, for example in Slovenia in later stages of the First World War or during the siege of Leningrad in 1941–44. Survival strategies extended beyond eating taboo foods and city dwellers took up small-scale home production along with foraging and purchase of food from peasants in the nearby countryside. Thus, one innovative strategy of coping with scarcity was effectively a return to former or traditional foodways. Another important strategy was barter and sharing within networks such as extended families, neighbourhoods or particular communities. Food shortages gave rise to extensive black markets which not only heightened inequalities in food consumption between different social groups, but also eroded conventional moral and ethical standards.[16]

Food and eating should not be understood as simply ingesting calories and nutrients, but rather as wider cultural and religious practices. German Jews were confronted with the challenge to uphold their dietary laws in wartime during the First World War and entitlements to food and eating were understood within the context of reciprocal relationships during the Spanish Civil War. Another problem was monotony which became a major grievance among British soldiers in the First World War and the difficulties in obtaining high-prized foods such as meat led to extensive discontent in Britain during the Second World War, although the country escaped the severe shortages and starvation common on the European continent.

---

15   Smith 2000.
16   Roodhouse 2006.

Thematic Structure of the Book

This book is divided into four parts, which examine the relationship between food and war in twentieth century Europe from a range of perspectives.

Part I: Soldiers and their Food

Part I explores the challenges involved in feeding the German and British armies during the First World War. Drawing on previously unused sources, these chapters enhance our understanding of army provisioning in the Great War. Lack of prewar planning precipitated experimentation and soldiers were forced to adapt to new eating patterns and unfamiliar foods, which raised particular problems for Jewish soldiers serving in the German army. Despite plentiful rations, soldiers' diets in Britain fell short of nutritional requirements and there is evidence of dietary deficiencies.

Chapter 2 by Peter Lummel examines wartime innovations in feeding the German Army. The country's internal struggle over supplies drove the quest for substitutes and innovative solutions to prevent hunger. Many 'civilian' soldiers found themselves eating previously untried foods such as factory-produced jam and tinned meat. The importance of vegetables in German Army rations provides a counterpoint to the British experience of army provisioning. This topic is discussed in Chapter 3 by Rachel Duffett, which challenges the dominant narrative that British Army provisioning on the Western Front was an unmitigated success. In reality, soldiers' rations, particularly those in the exposed areas of the frontline, were often deficient in quantity. The ration, which frequently consisted of tinned bully beef and hardtack biscuit, also lacked variety. This resulted in widespread health problems such as the gastric disorders, boils and bleeding gums. The final chapter in this part by Steven Schouten analyses the impact of the First World War on the kosher foodways of German Jews. Jewish soldiers were expected to adapt to a gentile diet and the majority ate pork, which was the meat commonly supplied to soldiers. This was perceived as an unavoidable aspect of a military existence, but Jewish soldiers also found ways to uphold their foodways, for example during holidays. Jewish women's attachment to kosher eating actually deepened despite food scarcity experienced on the home front. Food shortages, thus, had the effect of renewing a Jewish sense of identity.

Part II: Home Front: The Citizens Adapt

Given the priority generally attached to military requirements, the second part focuses on the difficulties experienced by civilians to secure adequate food supplies. These chapters draw attention to the role of women as food providers. Case studies of Germany and parts of the Austro-Hungarian Empire during the First World War examine the emergence of food consumption hierarchies and the rise of substitutes. Civilian responses have to be placed within a wider political framework. For instance, anti-Austrian and anti-German sentiment among the Czechs found expression in a popular outcry against the prospect of adding oak bark to bread flour. Civilians were forced to adapt to these altered circumstances and, as illustrated by the example of the Spanish Civil War, survival strategies have to be understood within the framework of social networks. The mental health implications of coping with extreme situations such as famine are explored in the context of the siege of Leningrad.

Chapter 5 by Hans-Jürgen Teuteberg explores the shortcomings of food control and inequalities in food provisioning in Germany. Civilian supplies were greatly reduced compared with prewar consumption and in the face of loopholes in the rationing system a huge black market developed. These circumstances resulted in a new food consumption hierarchy which privileged food producers, self-suppliers and highly paid war workers. By contrast, other city dwellers, families of soldiers and old people were most likely to suffer real hardship. Martin Franc's chapter traces the rise of natural and synthetic food substitutes which transformed staple foodstuffs such as bread and beer in Czech lands. Oak bark was never actually added to bread flour on a significant scale, but the experiments provided a focus for political discontent in the last years of the Austro-Hungarian Empire. Chapter 7 by Maja Godina Golija charts wartime life in Slovenia, which was dominated by a daily battle to obtain enough food to survive. Rations were not sufficient to avoid hunger, a flourishing black market developed and during the later war years there were frequent protests and food riots. An increased reliance on substitutes included not only wild plants but also dogs and cats as food for human consumption. The following chapter by Alicia Guidonet Riera analyses survival strategies in the face of intense food shortages during and after the Spanish Civil War from an anthropological perspective. On the basis of oral evidence, Guidonet highlights the key role of diverse supply channels and the wider social and cultural significance of reciprocal networks of assistance and exchange. Finally, Chapter 9 by Pavel Vasilyev draws attention to the rise of alimentary and pellagra psychoses in besieged Leningrad. Previously unexamined Soviet medical texts, which detail doctors' attempts to diagnose and classify mental breakdown caused by starvation provide a new perspective in the endeavour to reconstruct the features of everyday life in the besieged city.

Part III: Home Front: The State Intervenes

The state was the primary agency determining the allocation of food supplies. Part III examines food rationing systems and controls in Britain, France and the Netherlands during the Second World War. Wartime conditions altered the relationship between state and society. Extensive propaganda to encourage compliance had a limited effect, civilians frequently resisted official food policy and black markets were rife. The section juxtaposes the different food systems of belligerent and occupied countries. In Britain food shortages were not as severe as in France, but the worst food crisis in western Europe during the Second World War was the Dutch famine of 1944–45.

Chapter 10 by Ina Zweiniger-Bargielowska shows that food consumption hierarchies also emerged in Britain despite the fact that rationing was generally considered a great success. The policy of fair shares was compromised by inequalities of sacrifice between social classes, between men and women and official distribution channels were bypassed in the black market. British flat-rate rationing policy was supplemented by an expansion of communal feeding to allow for differential energy requirements and the chapter by Peter Atkins focuses on British Restaurants, which served mid-day meals to the general public. Atkins explores several issues, including regional variations, food served, customer attitudes, and their overall impact. Historians have not paid sufficient attention to the health consequences of the Second World War in France. Chapter 12 by Isabelle von Bueltzingsloewen highlights the demographic impact of the war demonstrated by rising mortality rates. The German imposed rationing system in occupied France did not provide enough food and undernutrition soon became a major problem. von Bueltzingsloewen examines how medical experts, by exposing the escalating health crisis, were able to challenge the occupation and to undermine the legitimacy of the Vichy regime. The next chapter by Kenneth Mouré analyzes the relationship between state controls to manage scarcity and responses to shortages in France by focusing on the black market. In the face of insufficient rations a large black market emerged, which further increased the defects in the state administered system. Finally, Chapter 14 by Adel den Hartog discusses nutrition education in the occupied Netherlands. The chapter focuses on the relationship between the Germans, the Dutch authorities, nutrition education workers and the population, paying particular attention to the nutritional message, food rations and ways of coping with food shortages. During the hunger winter of 1944–45, Dutch administrators were faced with an almost impossible task in their efforts to prevent mass starvation.

Part IV: War, Modernization and Innovation

The final part, which focuses on nutrition science, modernization and innovation, highlights the role of war in effecting longer-term changes. This is demonstrated both with regard to the rise of new foodstuffs and advances in food technology. Neutral countries such as Denmark and Iceland were not immune to the impact of war. Shortages and the disruption of agriculture resulted in a unique nutritional experiment in Denmark in the First World War and Iceland became a key provider and military base in the Allied war effort during the Second World War. Wartime shortages frequently precipitated the rise of substitutes. Commonly these fell out of use with the ending of hostilities, but one exception was horsemeat in France. The exigencies of war provided an incentive for technological innovation as illustrated in a case study of the impact of war on innovation in food technology Germany in the Second World War.

Chapter 15 by Svend Skafte Overgaard explores the culture of nutrition underpinning Denmark's rationing scheme, which was based on a low-protein diet advocated by the controversial reformist Dr Mikkel Hindhede. This was a great success and Hindhede subsequently proclaimed the rationing year of 1917–18 as 'The Year of Health' in which Denmark registered its lowest-ever mortality rate. Chapter 16 by Örn D. Jónsson and Guðmundur Jónsson discusses the impact of war on the modernization of the Icelandic diet in the twentieth century. Whereas the First World War precipitated a considerable reduction in the energy value of the Icelandic diet, the Second World War was a period of unprecedented prosperity. The next chapter by Alain Drouard explores the rise of horsemeat for human consumption in France which acquired prominence during the Franco-Prussian War of 1870–71 and remained popular for the best part of a century. The mechanization of agriculture after 1945 resulted in a decreased equine population, causing a rise in prices which precipitated the end of the widespread consumption of horsemeat in France. Finally, Chapter 18 by Ulrike Thoms discusses army food research and technological innovation during the Third Reich. Improved technologies with regard to dehydration and freezing as well as the development of substitutes helped to feed the German army during the war. In the longer term, the Nazi's massive investment in food technology provided West Germany's food processing industry with significant advantages which contributed towards the country's postwar economic recovery.

By providing a complex and nuanced understanding of the relationship between food and war in twentieth century Europe, this book demonstrates the importance of a comparative perspective. This is part of a growing body of scholarship but, unfortunately, there are still many gaps in the literature. This book, further, brings together a significant body of new research in a single volume. We hope that in the future scholars will explore countries and case studies which have as yet not received the attention they deserve.

References

Barnett, L.M. 1985. *British Food Policy during the First World War*. London: George Allen & Unwin.

Bonzon, T. 2006. Consumption and total warfare in Paris (1914–18), in *Food and Conflict in Europe in the Age of the Two World Wars*, edited by F. Trentmann and F. Just. Houndmills, Basingstoke: Palgrave Macmillan, 49–64.

Bonzon, T. and Davis, B. 1997, Feeding the cities, in *Capital cities at war: Paris, London, Berlin 1914–19*, edited by J. Winter and J.-L. Robert. Cambridge: Cambridge University Press, 305–41.

Collingham, L. 2011. *The Taste of War: World War Two and the Battle for Food*. London: Allen Lane.

Davis, B.J. 2000. *Home Fires Burning: Food, Politics, and Everyday Life in World War I Berlin*. Chapel Hill, NC: University of North Carolina Press.

Futselaar, R. 2006. The mystery of the dying Dutch: can micronutritent deficiencies explain the difference between Danish and Dutch wartime mortality?, in *Food and Conflict in Europe in the Age of the Two World Wars*, edited by F. Trentmann and F. Just. Houndmills, Basingstoke: Palgrave Macmillan, 193–222.

Hionidou, V. 2006. *Famine and Death in Occupied Greece, 1941–44*. Cambridge: Cambridge University Press.

Kamminga, H. 'Axes to grind': Popularizing the science of vitamins, 1920s and 1930s, in *Food, Science, Policy and Regulation in the Twentieth Century: International and Comparative Perspectives*, edited by D.F. Smith and J. Phillips. London: Routlege, 83–100.

Lih, L.T. 1990. *Bread and Authority in Russia, 1914–21*. Berkeley and Los Angeles, CA: University of California Press.

Moskoff, W. 1990. *The Bread of Affliction: The Food Supply in the USSR During World War II*. Cambridge: Cambridge University Press.

de Nijs, T. 2006. Food provision and food retailing in The Hague, 1914–30, in *Food and Conflict in Europe in the Age of the Two World Wars*, edited by F. Trentmann and F. Just. Houndmills, Basingstoke: Palgrave Macmillan, 65–87.

Nissen, M.R. 2006. Danish food production in the German war economy, in *Food and Conflict in Europe in the Age of the Two World Wars*, edited by F. Trentmann and F. Just. Houndmills, Basingstoke: Palgrave Macmillan, 172–92.

Olson, M. 1963. *The Economics of Wartime Shortage: A History of British Food Supplies in the Napoleonic War and in World Wars I and II*. Durham, NC: Duke University Press.

Offer, A. 1989. *The First World War: An Agrarian Interpretation*. Oxford: Clarendon Press.

Roodhouse, M. 2006. Popular morality and the black market in Britain, 1939–55, in *Food and Conflict in Europe in the Age of the Two World Wars*, edited by F. Trentmann and F. Just. Houndmills, Basingstoke: Palgrave Macmillan.

Schmidt, J. 2007. How to feed three million inhabitants: Berlin in the first years after the Second World War, 1945–48, in ICREFH IX, 63–73.

Smith, D.F. 1997. Nutrition Science in the two world wars, in *Nutrition in Britain: Science, Scientists and Politics in the Twentieth Century*, edited by D.F. Smith. London: Routledge, 142–65.

Smith, D.F. 2000. The rise and fall of the Scientific Food Committee during the Second World War, in *Food, Science, Policy and Regulation in the Twentieth Century: International and Comparative Perspectives*, edited in D.F. Smith and J. Phillips. London: Routlege, 101–16.

Teich, M. 1995. Science and food during the Great War: Britain and Germany, in *The Science and Culture of Nutrition, 1840–1940*, edited by H. Kamminga and A. Cunningham. Amsterdam: Rodopi, 213–34.

Treitel, C. 2008. Max Rubner and the biopolitics of rational nutrition. *Central European History*, 41, 1–25.

Trentmann, F. and Just, F. ed. 2006. *Food and Conflict in Europe in the Age of the Two World Wars*. Houndmills, Basingstoke: Palgrave Macmillan.

Trienekens, G. 2000. The food supply in The Netherlands during the Second World War, in *Food, Science, Policy and Regulation in the Twentieth Century:*

International and Comparative Perspectives, edited by D.F. Smith and J. Phillips. London: Routlege, 117–33.

Vernon, J. 2007. *Hunger: A Modern History*. Cambridge, MA: Harvard University Press.

Vincent, C.P. 1985. *The Politics of Hunger: The Allied Blockade of Germany, 1915–19*. Athens, OH: Ohio University Press.

Wall, R., and Winter, J. ed. 1988. *The Upheaval of War: Family, Work and Welfare in Europe, 1914–18*. Cambridge: Cambridge University Press.

Weatherall, M. 1995. Bread and newspapers: the making of 'a revolution in the science of food', in *The Science and Culture of Nutrition, 1840–1940*, edited by H. Kamminga and A. Cunningham. Amsterdam: Rodopi, 179–212.

Zweiniger-Bargielowska, I. 2000. *Austerity in Britain: Rationing, Controls and Consumption, 1939–55*. Oxford: Oxford University Press.

PART I
Soldiers and their Food

Chapter 2

Food Provisioning in the German Army of the First World War

Peter Lummel

Introduction

The First World War has been described as 'the great seminal catastrophe of the twentieth century', an event which initiated an epoch of great changes.[1] The years before might be described as a first Cold War, typified as they were by mutual distrust, the arms race and propaganda. Such negative factors determined the relationship between the great powers of Europe in the age of imperialism, with Germany, Austria-Hungary and Italy in conflict with France, Great Britain and Russia. Germany was an economic world power on the eve of the First World War, striving for expansion and recognition; it was a leading industrial nation with worldwide exports. The industrial production of food and semi-luxury goods took place on a wide scale, and railways, steamships, central market halls, co-ops, chains of retail shops, department stores and mail order companies were all evidence of a modern trading nation.[2] Widespread and significant hunger was not a feature of German life in the years preceding 1914.

When Germany entered the war following the invasion of Luxembourg and Belgium on 1 August 1914, there was a conviction that it would be a short and, for them, successful conflict. The military strategy of the *Schlieffenplan* was intended to defeat France rapidly in order to be able to then wage war on Russia in the East, but the course of war proved to be completely different. Germany entangled itself in fighting in many different theatres and the long-lasting static warfare on the Western Front resulted in immense losses on both sides. The British naval blockade cut Germany off from imports of industrial raw materials, food and other goods of daily use. When the Americans entered the war in 1917, because of their military strength and economic resources, Germany could no longer anticipate success in the conflict.

The industrialized and de-personalized nature of the war was unimaginable for those caught up in its horrors. New highly developed mass-produced weapons

---

1 The term was employed by the American historian and diplomat George F. Kennan. See Neugebauer 2007: 4.

2 Hans Jürgen Teuteberg has explored these questions extensively, see Teuteberg 2004.

were used at the front: machine-guns, flame-throwers, mines, poison gas, artillery shells, tanks, airships and aeroplanes. The type of warfare characterized by the strategies and military formations employed in the Franco-Prussian War of 1870–71 had disappeared. Soldiers found themselves in a war of attrition; they lay in dugouts and fought against the damp and cold, as well as diseases and the terrifying anonymity of industrial weaponry, which brought death and mutilation for millions of people.[3] The nature of the war resulted in the provisioning of the German Army reaching a scale that had never been previously experienced.

Nevertheless, until now there has not been any critical scientific research into the military's food supply in the First World War. Existing research on the war concentrates primarily on the military and political aspects, and less frequently on its economic and social questions.[4] The question of provisioning the army has only been considered from the medical viewpoint, for example in the detailed work edited by the former President of the Imperial Public Health Department, Franz Bumm 1926–8, *Germany's State of Health under the Influence of the World War*, which included articles by Max Rubner and other well known scientists.[5] In National Socialist Germany, the feeding of soldiers in the First World War was examined frequently by army doctors as well as university scholars in both medical and medical history research. A key study was written by Walther Kittel, Walter Schreiber and Wilhelm Ziegelmayer in 1939 and a medical thesis was published by Friedrich Fernow in 1937 that explored the relationship between war, food and disease.[6] Christoph Feldmann wrote an unpublished state examination work in 1995 under Hans Jürgen Teuteberg's supervision on the topic 'The development of armed forces provisioning in the 19th and 20th century', however, he did not consider the experience of the First World War.[7]

This chapter uses new source material and literature to evaluate the records of the Military History Research Institute of the Federal Armed Forces in Potsdam. Papers from the Imperial War Nutrition Office, military prescriptions, war memoirs, studies and autobiographies of protagonists of the food supply bureaucracy, physical and medical evaluations, economic analyses and other studies have all helped to illuminate this complex topic. The main questions considered in this chapter are: what were the daily ration scales and logistics of German army provisioning? How did the delivery match the plans and were the food supplies adequate? Were there any pioneering food innovations in, or resulting from, the army's supply procedures? And finally, did the experience of the First World War have a long term impact on Germany's food systems?

---

3   Neugebauer 2007: 30–36, Berghan 2009, Hirschfeld et al. 2003, Hirschfeld et al. 2006.
4   See Broadberry, Harrison 2006, Neugebauer 2007.
5   Bumm 1928.
6   Kittel et al. 1939, Fernow 1937.
7   Feldmann 1995.

Food Rations in the Army

At the beginning of the First World War soldiers' daily iron ration consisted of 200 grams of zwieback (hard-baked bread rather like hardtack biscuit), 200 grams of tinned meat 150 grams of preserved vegetables and 25 grams each of salt and coffee.[8] In addition, the men received regular daily rations of bread, meat, vegetables, spices and hot drinks. In 1914 the soldiers received 750 grams of bread, 375 grams of fresh, salted or frozen meat, 125 grams of rice, pearl barley or grits as well as 25 grams of salt and coffee. German soldiers operating in the frontline received at least 3,200 calories per day at the start of the war, and this included 117 grams of protein.[9] The daily rations depended upon availability and officers' decisions. For example, a General might increase the meat and vegetable portions by up to one third or distribute special semi-luxury items such as cheese before combat operations or after special endeavours by the soldiers.[10]

In 1914, the active service rations were not too far distant from the basic foodstuffs of the preceding 50 years. Recently developed items from the preserved vegetables industry, and new corn and potato products had not yet become a key part of the soldiers' food. However, a mere two years later in 1916 the spectrum of distributed food had been greatly extended, not only as a result of food shortages, particularly those of meat and fat but also because of the demand from the soldiers themselves for greater variety.[11] Changes included fish as an alternative to meat. This could be fresh, salted, smoked, or marinated in sauces in tins. Dried potato flakes, a new product, were made available to supplement fresh potatoes, which often reached the troops in far less than the requested quantity because of their perishable nature. The vegetable supply was extended by carrots, swedes, turnips, kohlrabi, string beans and spinach. Cabbage, savoy, white, green and red, were all regularly distributed, supplemented with noodles and dried fruit. For the fat ration, butter, lard and tinned pork were provided. Spreads were increasingly allotted in the form of soft sausages and cheese, due to the shortage of jam (also called 'hero fat').[12] Extras such as sweets and other dainties from the homeland, appeared in the early years of the war, but became rarer as the war continued.

As the war progressed, the rations assigned to the army provided a stark contrast to those on the home front as the food supply worsened dramatically. A document published by the Imperial War Nutrition Office highlighted the difficulties of the situation by using the example that the daily portions of fat and meat in the army were the same as the city-dwellers' weekly ration in the winter of 1916–17.[13] It was

---

8 Tinned food became a central element of the iron ration from 1880 onwards; see Schneider 1910: 25–35.

9 Kittel et al. 1939: 11.

10 Wedel von 1915: 151, Gesche 1926: 46.

11 Ibid.

12 Hartmann 1917: 7.

13 Hartmann 1917: 22.

only towards the end of war that the provisioning of the armed services, particularly the navy, suffered in a similar fashion to the civilians at home.[14] By 1917–18, the civilian bread ration was regarded as derisory, although that of the fighting forces was maintained at a far higher level. Vegetables were distributed primarily in the dried form. The salted vegetables that were available were regarded as having an inferior taste; the supply of legumes also decreased significantly. Imported food from other continents virtually disappeared and as choice declined, the diet became far more monotonous. These privations did not appear to apply to the military elites, and towards the end of war, luxurious dishes served to officers in the casinos and other such places far away from frontline were criticized.[15]

In addition to an overview of the German Army's rations during the First World War, the table below provides a comparison of the daily ration of German and British soldiers, demonstrating not only issues of availability but also those of food preferences in the two nations.[16] While German soldiers received a lot of bread, fat, vegetables and a wide variety of wheat products, British troops had considerably more meat and bacon in their daily diet. Tea and sugar were more plentiful in the British Army, as coffee was rarely drunk by British soldiers who were also less interested in spices for their food, of which there was a large variety in the German Army.

Logistics

While the Central Department for Procuring Military Food Supply managed the bulk of the nation's supplies, with food storage sites inside Germany and special supply departments near the border, the responsibility for food procurement shifted to those based outside the country. Commissioners or *Intendanten*, were responsible for food provisioning of soldiers in the war zones. In line with army hierarchy these were subdivided into general, army, and corps as well as division commissioners, each with their associated commissariats. The challenge of food supply had been relatively insignificant in earlier conflicts both because of their brevity and the much smaller armies involved. Although Wallenstein's method – feeding the war by the war, through pillage as the army advanced – was still considered the primary strategy as late as 1910 it could not function in a conflict on the scale of the First World War.[17] Newer methods were required to feed so many static soldiers, such as the confiscation of food or its purchase from local farmers and retailers, often financed by taxes collected from enemy territory.

---

14   Gesche 1926: 76, see also Merkel and Fikentscher in Bumm 1928: 163, 215 regarding the navy.

15   Gesche 1926: 88.

16   See Chapter 3 by Rachel Duffett.

17   Schneider 1910: 35, Wedel von 1915: 154, Hartmann 1917: 1, Feldmann 1995: 12.

Table 2.1   German and British Armies' Food Ration in 1916*

Food	German Army	British Army
BREAD	750g of bread	397g of bread
MEAT	250g of a fresh/spicy/frozen meat or 150g of smoked meat/bacon/salami/tinned meat	340g of meat (often bully beef) and 93g of bacon
FISH	as meat substitute 250g – 600g fish	as an occasional meat substitute
VEGETABLES	1500g of potatoes/1200g of beets/savoy cabbage or 450g of sauerkraut/400g of spicy spinach from barrels or 300g of dehydrated potatoes/250g of legumes/potato flakes/ spicy cut beans or 150g of preserved vegetables or 125g of dried fruit	227g of vegetables (very often only potatoes)
CORN	instead of vegetables alternatively 125g rice, grains of pearl barley or grits alternatively: 125g of semolina or 250g of flour or 200g of pasta	57g of rice
FAT/SPREADS	65g of butter/lard/fat pork in tins alternatively: 125g of fruit jam or cheese/125g of cooked pork and beef sausage, blood and liver tinned sausages	28g of butter 85g of fruit jam 57g of cheese
SUGAR	none (only with tea 17g)	57g
SPICES	25g of salt, 2.5g of mustard, 0.1g of peppers, 25g of onions, 0.4g of pepper, 2g of caraway, 0.1g of clove, 0.05g bay leaves, 0.2g of marjoram, 0.05 l of vinegar, 0.05 l of salad oil. 3g of cinnamon	salt, mustard and pepper
HOT DRINKS	3g of tea or 25g of distilled coffee	14g of tea with 28g of condensed milk

*   Hartmann 1917: 22. See also Chapter 3 by Rachel Duffett. N.B. the quantities in the table are those for the troops in the reserve lines, those on the front line would have received more food.

The supplies required for the millions in the army were immense. The documents of the Imperial War Nutrition Office indicate that in the first two years of war alone, almost 1.2 billion kilos of flour, one million cattle, one million pigs and 570,000 sheep as well as 275 million kilos of canned meat were delivered to the army from Germany.[18] An astonishing quantity of semi-luxury items were also transported to the army. This includes 62.5 million kilos of coffee, 15 million kilos of cocoa and tea, 2.7 million hectolitres of beer and 8.5 billion cigars and cigarettes. In total, during the first two years of the war, 8 billion kilos of food and semi-luxury items were supplied from Germany to the troops serving abroad.

In the context of mobile warfare, particularly in the East, and the great size of the combat zones, these huge quantities of supplies could only be delivered at the rate required through the use of a modern railway system.[19] The armed forces had priority use of the trains for troops and supply movement. Large field base stores were established in enemy territory at the end of railway lines. Where new track was required to complete the system, units of up to 3,000 workers could build 3 kilometres per day.[20] The slower ship transports which were also used to stock the large field stores were of minor importance in comparison to the freight carried on the railways. In foreign countries, base stores or *Etappenmagazine* were often opened far away from the fighting front for security reasons. From there the goods were driven hundreds of kilometres up to the field base stores (*Feldmagazine*), and then distributed to the fighting troops nearby in convoys of horses and carts or lorries. A convoy could transport up to 54,000 kilos of reinforcements effectively taking over the function of a mobile magazine. The daily rations were collected by the troops from the field base stores mainly at night because of the danger of attracting enemy attention.[21]

In order to guarantee the troops' supply, food processing enterprises were taken over or set up under military supervision in the occupied areas, for example, grain mills, bakeries, breweries and slaughterhouses, drying facilities, fruit processing companies, and mineral water factories.[22] In addition, basic foodstuffs such as bread, meat and sausages were provided by military organizations.[23] Transported cattle were slaughtered in simple mobile constructions, and bread was made in mobile ovens. Travelling bakeries had been in use since 1896, but these were now adapted to undertake a 24 hour continuous operation which could produce 1,920 loaves daily. Field kitchens had also been developed following the experiences of the Russo-Japanese War.[24] The German company Senking had won a competition in cooperation with the company Magirus and following testing in 1907, 1,000

18   Hartmann 1917: 4.
19   Schneider 1910: 36–47, Großer Generalstab 1913: 300, Hartmann 1917: 4, 29.
20   Schneider 1910: 36.
21   See descriptions in letters from the frontline Hirschfeld 2006: 93.
22   Schneider 1910: 1.
23   Schneider 1910: 129–40, Hartmann 1917: 34, Feldmann 1995: 40–47.
24   For field kitchens see Schneider 1910: 77, Wedel von 1915: 157.

horse-drawn field kitchens were available in 1914, a number which had increased to 4,000 by the end of war. The technology meant that it was possible to cook while marching: legumes, beef and mutton needed approximately two hours to cook in 200 litre pots, and potatoes, rice and pork needed around one hour. At the same time that the food was cooking, it was possible to boil 70 litres of coffee in a second pot. Field kitchens were particularly suitable for the production of soups and stews, in which dried as well as tinned foods were used together with fresh products. Field kitchens made it possible to provide the men with hot food each day, and at the same time the food supply became more efficient and streamlined with a consequent decrease in waste.

Evaluation of the Food Supply

Provisioning the army in the First World War was an unprecedented challenge; the numbers of men involved were up to 20 times greater than those of the Franco-Prussian War. The demands of modern warfare with its new weapon systems, the frequency and duration of battle and the physical as well as mental strains were not comparable with earlier times. Additionally, it was not only scientific knowledge but also the eating habits of the soldiers which had changed considerably over the decades before the war. The evaluation of the preparation, acquisition and quality of food, the reliability of provisioning and effective use of resources, the legal and organizational changes required, and the technical and personnel needs had to be undertaken in the light of the complex requirements of this new style of warfare.

In a new departure for Germany, food was rationed at a governmental level in an effort to distribute it in sufficient quantities to the population, although these became increasingly meagre as the war progressed.[25] After initial difficulties, it proved that the centralized management of the supplies required from the home front by the Prussian War Office was the correct decision and the use of supply officers operating in the field was equally successful. The daily rations for soldiers were protected in such a way to ensure that no soldier experienced the hunger felt by German civilians and investment in modern ovens and field kitchens also helped to deliver good food to the men. Factories were established through subsidies from the Ministry of Finance for the production of tinned and preserved foods and dried vegetables for the army and food was also purchased for the troops from local bakers, butchers and grocers.

There were also significant omissions and a number of false assumptions. The failure with the most serious consequences was that, despite Germany's attention to matters of armament, it entered the war without any proper preparation in the matter of economic policy.[26] The state had no mechanism for collecting data on national food production and stores, nor did it have a plan for the organization of

---

25   See Chapter 5 by Hans-Jürgen Teuteberg.
26   Verified by Delbrück 1924, Skalweit 1927: 5, Bumm 1928: 3.

the food economy in time of war. An additional key difficulty was the quality of the labour employed in the whole of the army's provisioning processes, from the stores to the field kitchens. These tasks were increasingly performed by unqualified staff or prisoners of war, as healthy men were required at the front.[27] Problems of supply and food shortages in Germany were exacerbated by theft; looting from the stores on the home front began in 1915 and from 1917 onwards the reports of train lootings by German soldiers multiplied.[28] In addition, during the last two years of the war there were numerous instances of fraud, when military units purposely over-stated the number of field rations required.

A fundamental problem was the assumption by the German state that it was possible to deal with the military and civilian food needs as if they were unconnected. The preferential treatment given to the army led to disastrous hunger on the home front, beginning in the winter 1916–17. It was evident that many of the soldiers who joined up towards the end of war had constitutions that had already suffered as a result of the privations in Germany.[29] In 1918 the army consumed 30 per cent of the total bread grain and 60 per cent of the total cattle and pork available from German production, an indication that the military had not taken into account the latest findings of the leading German nutritionist Max Rubner. He had investigated the connection between the degree of physical work and the caloric intake required for its performance. Rubner's deductions were that the level of meat protein in the army rations was substantially too high and that alternative foods could be substituted.[30]

A critical assessment of the feeding of an army of millions should not ignore the challenges that the completely new requirements in the development of the supply chain and the men's own dietary preferences presented. Scientific knowledge, technical achievements and new logistical solutions ensured that both hunger and deficiency diseases were avoided in the field. However, it is as yet unclear why little importance was attached to the food question before 1914, despite a few warning voices amongst political and military figures. The question still remains as to why a country of Germany's economic sophistication had not considered the issues of food stores and supply in view of the likelihood of a blockade.[31] It appears that deficiencies in planning, incorrect conclusions and a lack of interest and support amongst the political and military elites led to the failures of food policy and their serious consequences.

---

27   Roland 1917, Merkel in Bumm 1928: 163, Kittel et al. 1939: 93, 196, also soldiers' letters in Hirschfeld 2006.

28   Gesche 1926: 76.

29   Rubner in Bumm 1928: 20–41 verifies the loss of weight which was diagnosed during the First World War in German institutions: men lost 33 per cent on average and women 25 per cent.

30   ICREFH IX: 4.

31   This was one of the key differences compared with the Second World War, see Chapter 18 by Ulrike Thoms.

Product Innovations in Army Provisioning

The German food industry actively sought new products during the war years.[32] Concerns about shelf-life as well as the desire for artificial foods were important in terms of military supply.[33] Although the development of new artificial foods was rare during the war, artificial yeast, a food abundant in proteins but cultivated using purely mineral nutrients, was an important exception. Numerous substitute foodstuffs had already been developed before 1914: margarine, coffee substitutes, bouillon cubes, artificial sweeteners and artificial honey being the key examples. The war fuelled the search for new food products and product variants, and these were developed in huge varieties: around 11,000 products were registered officially and considerably more may have existed. Most of these substitutes were not for the armed forces but for the civilian market, where hunger and high demand resulted in inferior products and adulterated food entering the market.

Although there was no lack of innovation in the food industry, revolutionary changes seem to have been rare; exceptions were artificial yeast, dried potato flakes, industrially produced jams and sunflower oil, all products which proved their usefulness through their broad range of application in military provisioning. Clearly, the war-induced lack of raw materials was a hindrance for innovation. The shortages of suitable tin as well as that of specialist staff prevented the canning industry from producing better and more palatable products. The dehydrated food industry failed to provide a wide range of new vegetable products, and tests for quick-freezing fish, meat, fruit and vegetables did not achieve satisfactory results.[34]

Changes to the Food System Resulting from the First World War

It appears that the First World War did not have a long-lasting impact on Germany's food system. Max Rubner wrote in 1928

> It was claimed during the war that its complete reversal of the diet would continue into peacetime. This did not happen.. Meal forms have always more and more approached that of the earlier pre-war food. The traditional food is so deeply established in people's consciousness that the return to the old ways should not surprise.[35]

The observation of Rubner seems to be clear: continuation of traditional foods and ways of eating rather than changes born out of the deprivations of the war. Another

---

32   For the aspect of war and innovations in general, Boot 2006, Auerbach 1920: 502–14, Rubner in Bumm 1928: 3. See Chapter 5 by Martin Franc.

33   Kerp in Bumm 1928: 77–122.

34   Spiekermann 1997: 33.

35   Rubner in Bumm 1928: 40.

article in the same book by Doctor Kerp on 'Supplies of substitute food' indicates only too well why the products hastily constituted in wartime rarely maintained their position in the retail trade once peace had resumed.[36]

During the First World War, the state had no strategy for dealing with the influx of new and often dubious substitute foods that appeared, other than to proceed reactively with regulations, announcements, controls and threats of punishment, usually with little confidence in their effect.[37] At the same time, the pressure on the legislature and executive to protect the health of consumers grew. It continued until 1918, when in the *Announcement of 18th of April 1918* substitute food was, for the first time, defined by law and its producers required authorization and registration. Subsequently the food industry was compelled to provide full and clear labelling of its products, including: name, company, date of manufacture and contents. Consumer interests now had to be taken into account on a whole new scale. In 1927, existing German food legislation dating back to1879 was replaced with new law that reflected the gains made in consumer protection following the negative experiences of the First World War. The position of consumers improved considerably as a result of the conflict, in a manner that still has consequences for contemporary society. Thus, when Max Rubner judged that consumers preferred to return to their pre-war foodways, he could have not included weak legal protection in his assessment.

The First World War did change certain German eating habits. Before 1914 there were wide variances between town and country preferences and even greater regional differences.[38] Soldiers fighting together in the army came from all over the nation and inevitably food habits were adjusted through contact with unfamiliar preferences and foodstuffs. During the war some men ate factory jam, wet saltwater fish and tinned foods for the first time and there are clear indications that certain food items gained acceptance in these years. The war resulted in soldiers being offered a considerably wider selection of vegetables than that which they would have eaten in peacetime, given the restrictions of regional consumption patterns.[39] While industrially produced dried vegetables were often refused by soldiers, due in part to their poor quality, there were no specific complaints about fresh vegetables.

Similarly, the civilian population had rediscovered vegetables following the severe hunger experienced, as evidenced by the allotment boom in German cities after the war. A comparison of the years 1912–13 and 1927 demonstrates that the number of allotments in Berlin grew from 1,841 to 138,000, in Hamburg from 253 to 35,269 and in Cologne from 578 to 36,900.[40] It could be assumed that

---

36   Kerp in Bumm 1928: 77–122.

37   See the annual reports of the Imperial public health department from 1914 to 1918.

38   Lesniczak 2003.

39   For changes in German vegetable consumption, see Teuteberg and Wiegelmann 1986: 257.

40   Spiekermann 1997: 37.

the common eating experiences of soldiers changed opinions about vegetable consumption, and from this time on vegetables became an important constituent of a meal, a change that was also supported by new scientific information of their nutritional value. A similar effect is indicated in the wider postwar consumption of fish. Soldiers' experiences of army food, of which fish was a regular part from 1916 onwards, shaped their civilian dining after 1918. Before the war, fish dishes were only common in coastal areas and in certain regions, but after the First World War fish meals became common, particularly for the Friday meal, mainly due to its valuable protein content.

Cocoa mixed with sugar was given to all fighting troops at the front as a pleasing semi-luxury item, and in the first two years of the war 15 million kilos of cocoa were distributed.[41] By the end of the war every soldier had been exposed to the pleasures of cocoa and chocolate, and as soon as the import of raw materials resumed, Germany saw an astonishing increase of their production and consumption, despite the extreme economic difficulty and inflation of that period. Just as smoking cigars and cigarettes virtually became a badge of the war generation, so cocoa and chocolate were established as easily transportable, filling, semi-luxury items across all social classes through the consumption habits of the armed forces. This widespread enthusiasm is indirectly confirmed by the orientation of the chocolate industry's advertising campaigns of the 1920s towards the broader population.[42]

Conclusion

The economic policy of Germany during the First World War was in many ways ineffective. However, the food provisioning of the German Army was sufficient in quantity throughout the war, not least because of good logistics, modern equipment and a complex infrastructure of store depots, but also because civilians starved to ensure that the soldiers ate. Supply of the army reflected its soldiers' demands for variety, and interestingly, vegetables played an important role compared with their use in the British Army. The quality of food was criticized by the men, in particular because of the poor quality of tinned foods and dried vegetables, the spoiling of foodstuffs during transportation and the poor food produced by the cooks. The war did not lead to any truly pioneering food innovations, but hunger as well as the more positive food experiences of the First World War led to certain long term changes in the German food system.

---

41   Hartmann 1917: 4.
42   Lummel: 2005.

References

Auerbach, F. 1920. Technische Errungenschaften der Lebensmittelgewerbe [Technical achievements in the food business], in *Die Technik im Weltkriege* [Technology in the World War], edited by M. Schwarte. Berlin: Mittler.

Berghan, V.R. 2003. *Der Erste Weltkrieg* [World War I]. Munich: C.H. Beck.

Broadberry, S. and Harrison, M. eds 2005. *The Economics of World War I*. Cambridge: Cambridge University Press.

Bumm, F. ed. 1928. *Deutschlands Gesundheitsverhältnisse unter dem Einfluss des Weltkrieges* [Germany's State of Health under the Influence of World War]. Stuttgart: Deutsche Verlagsanstalt.

Von Delbrück, C. with von Delbrück, J. 1924. *Die wirtschaftliche Mobilmachung in Deutschland 1914* [Economic Mobilization in Germany in 1914]. Munich: Verlag für Kulturpolitik.

Feldmann, C. 1995. Die Entwicklung der Militärverpflegung im 19. und 20 Jahrhundert [The Development of Armed Forces Provisioning in the 19th and the 20th century]. Unpublished script, Münster.

Fernow, F. 1937. *Über Truppenernährung in den Feldzügen 1740 bis 1918* [War Provisioning from 1740 to 1918]. Self-published.

Gesche, P. 1926. *Heeresverpflegung und Zusammenbruch* [Army Provisioning and its Collapse]. Kassel: self-published.

Großer Generalstab Kriegsgeschichtliche Abteilung ed. 1913. *Heeresverpflegung* [Army Provisioning]. Berlin: Mittler.

Hartmann, F. 1917. Die Heeresverpflegung [Food Provisioning in the Army], in Volkswirtschaftliche Abteilung des Kriegsernährungsamts. Beiträge zur Kriegswirtschaft, Heft 11.

Hirschfeld, G., Krumreich, G., and Renz, I. eds 2003. *Enzyklopädie Erster Weltkrieg* [Encyclopaedia of the First World War]. Paderborn: Schöningh.

Hirschfeld, G., Krumreich, G., and Renz, I. eds 2006. *Die Deutschen an der Somme 1914–1918. Krieg, Besatzung, verbrannte Erde* [Germans on the Somme 1914–1918]. Essen: Klartext Verlag.

Kittel, W., Schreiber, W. and Ziegelmayer, W. 1939. *Soldatenernährung und Gemeinschaftsverpflegung* [Soldier's Diet and Canteen Meals]. Dresden: Steinkopf.

Lesniczak, P. 2003. *Alte Landschaftsküchen im Sog der Modernisierung. Studien zu einer Ernährungsgeographie in Deutschland zwischen 1860 und 1930* [Old Regional Cooking 1860–1930]. Stuttgart: Franz Steiner Verlag.

Lummel, P. 2005. The images of chocolate and coffee in the mass media in Germany during the 20th century, in ICREFH VIII.

Neugebauer, K.-V. ed. 2007. *Grundkurs Deutscher Militärgeschichte, 2, Das Zeitalter der Weltkriege 1914 – 1918. Völker in Waffen* [German Military History: The Age of World Wars]. Munich: Oldenbourg.

Roland, J. 1918. *Unsere Lebensmittel. Ihr Wesen, ihre Veränderungen und Konservierung vom ernährungsphysiologischen und volkswirtschaftlichen Standpunkt* [Our food]. Dresden and Leipzig: Steinkopf.

Schneider, P. 1910. *Die Verpflegung des Feldheeres* [The Provisioning of the Army]. Berlin: Mittler.

Skalweit, A. 1927. *Die Deutsche Kriegsernährungswirtschaft. Wirtschafts- und Sozialgeschichte des Weltkrieges* [The German Wartime Food Economy]. Stuttgart: Deutsche Verlagsanstalt.

Spiekermann, U. 1997. Zeitensprünge: Lebensmittelkonservierung zwischen Industrie und Haushalt 1880–1940 [Time transfers: Food conservation between industries and household 1880–1940], in *Ernährungskultur im Wandel der Zeiten*. Mühlheim/Ruhr, 30–42.

Teuteberg, H.-J. 2004. *Die Revolution am Esstisch. Neue Studien zur Nahrungskultur im 19. / 20. Jahrhundert* [Revolution at the Table. New studies of Culinary Culture in the 19th and 20th century]. Stuttgart: Franz Steiner Verlag.

Teuteberg, H.-J., and Wiegelmann, G. eds 1986. *Unsere tägliche Kost* [Our Daily Food]. Münster: Steiner.

von Wedel. 1915. *Von Wedels Offizier-Taschenbuch für Manöver, Übungsritte, Feldgebrauch, Kriegsspiel, taktische Arbeiten* [Pocket book for Manoeuvres, Riding Exercises, Field Use, War Games and tactical Work]. New edition edited by W. Balck, Generalmajor und Inspekteur der Feldtelegraphie. Berlin: Eisenschmidt.

Chapter 3

British Army Provisioning on the Western Front, 1914–1918

Rachel Duffett

Introduction

In 1931, Colonel R.H. Beadon, a veteran of the First World War, published a study of the British Army's supply systems during the conflict in which he quoted an earlier German strategist: 'the maxim of von Moltke that "no army food is too expensive" was faithfully observed throughout'.[1] Beadon was convinced that every effort, whether financial or logistical, had been taken to ensure that the soldiers received regular, good quality, rations. Beadon's analysis of the army's achievements was enthusiastic and uncritical, an approach reflected in the military's own assessment in its postwar review.[2] Subsequent commentators have agreed, their views summed up by John Burnett who claimed that 'for millions of soldiers ... wartime rations represented a higher standard of feeding than they had ever known before.'[3]

The dissenting voice amongst the general approval comes from the soldiers themselves, whose letters, diaries, and memoirs do not echo the official enthusiasm for their rations. Investigations are hampered by a lack of documentation; the official records of provisioning have not been accorded the longevity of other documents from the conflict and the day-to-day details of food supply have not been preserved. Therefore, the narrative of army provisioning must be pieced together from the fragments that remain, including ration scales, cookery manuals, the gross statistics of the postwar review and the men's own accounts. This chapter will provide an overview of the practical challenges of provisioning rank and file soldiers on the Western Front and consider some of the problems of supply encountered by the army and their men.

---

1   Beadon 1931: 95.
2   *Statistics of Military Effort* 1999.
3   Burnett 1979: 271. Most recently, Gregory 2008: 282 has restated the consensus: 'Many perhaps most, soldiers were better fed in the Army than they had ever been in civilian life'. For a reappraisal of British army provisioning see Duffett 2011.

Pre-First World War Rations

The rations scales of the First World War were indicative of the efforts made to improve military provisioning in the sixty years preceding its outbreak. Army food had traditionally been poor. Research at Edinburgh University in the mid-nineteenth century had demonstrated that the diet of the prisoners in Perth Gaol was more nutritious than that of British soldiers.[4] Failures in the ration had been brought to the fore during the Crimean War, where the food supplied was inadequate, resulting in large numbers of soldiers being admitted to Scutari Hospital suffering from scurvy.[5] A series of government commissions ensued that, along with other areas of reform, strove to upgrade the men's diet through the improvement of the quality and variety of the foodstuffs offered and greater attention to catering practices. Food preparation had always been a haphazard affair. It was only in 1870 that the army had established its first School of Cooking in an effort to raise the standards in the cookhouse through formal training. Subsequently, in the *1889 Queen's Regulations and Orders*, it was made compulsory for each battalion to send Sergeant Cooks to the school, in order that they could supervise and educate other cooks in their units.[6]

Regardless of advancements in military feeding during this period, civilian nutrition for many remained poor. The Boer War recruitment process at the beginning of the twentieth century demonstrated that much of the British working class was under-nourished, given that approximately one third of volunteers were deemed unfit to serve, mainly because of factors related to poor diet.[7] Major-General Sir Frederick Maurice, took a particularly bleak view of the experience, commenting that it demonstrated that the physical state of the nation was 'a far more deadly peril than any that was presented by the most anxious period of the South African War'.[8] The stature and health of many of the working-class volunteers reflected a diet that relied heavily on bread and jam, rather than bone-building fats and proteins. Indeed, the prospect of regular and plentiful food may have been an incentive for some of the volunteers, an opportunity to satisfy a hunger that was as much physical as it was patriotic. Certainly for the hungrier men, the First World War Recruiting Sergeants' promise to the inhabitants of the Salford slums of the unparalleled treat of 'meat every day!' was seductive.[9] Popular recruiting songs extolled the virtues of the military diet:

---

4   Smurthwaite 1993: 81.

5   Drummond and Wilbraham 1958: 395.

6   Cole 1984: 34.

7   For example, of the 3,600 men who applied to enlist at York, Leeds and Sheffield between 1897 and 1901, nearly half of them failed the medical examination on the grounds of 'under-development ... defective vision ... and decayed teeth'. Dwork 1987: 15.

8   Floud et al. 1990: 306.

9   Roberts 1971: 189.

Come on and join Lord Kitchener's army
Ten bob a week, plenty of grub to eat[10]

The First World War ration scales appeared generous at nearly 4,200 calories per day at the front. This figure has remained relatively unchanged as the British Army currently feeds its troops 4,100 in the field. The scales reflected the limitations of nutritional science at the beginning of the twentieth century. The key dietary building blocks of fats, carbohydrate and protein had been identified earlier, but it was not until the final decade of the nineteenth century that an understanding of calories emerged when Wilbur Atwater constructed a calorimeter. Atwater's work proved the new thermodynamic model that was to inform the military's approach to feeding: food as fuel for the human motor. Atwater's research had also made it possible to calculate the energy values of foods. However, biochemical knowledge remained imprecise and whilst Casimir Funk first coined the word 'vitamine' in 1912, when he determined that trace elements must be an essential part of a healthy diet, full understanding was yet to come.

Details of the Ration

During the conflict, food for the British soldier varied according to rank, chronology and position: officers, with their separate messes and greater resources, ate better than their men; the ration scales were reduced at certain points over the course of the war; and soldiers in the frontline trenches received more than those in the lines of communication behind. The ration was founded upon bread and meat, the mainstay of soldiers' food since it was first fixed at 1lb of bread and 12oz of meat per day in the Army Regulations of 1813.[11] The details of the daily scales for the last two years of the war are shown below in table 3.1.

In addition to the basic items, salt, mustard and pepper were issued to season the diet. Reductions had been made in several items from the levels at the outset of the war, including the 1 lb of fresh meat for frontline soldiers noted above which had stood at 1¼ lb in 1914. The army claimed that such savings were achieved 'without detriment to the adequate feeding of the soldier'. Economies were attributed to better training of cooks and greater efficiency in the inspections carried out by the Quartermaster-General's Services. Veterans reading those lines may not have recognized the generous appraisal of their provisioning. No doubt their attention would have been drawn to the next sentence in the review, where the army congratulated itself upon the 'millions of pound per annum' saved by the reduction in the ration. [12]

---

10   Arthur 2001: 3.
11   Skelley 1977: 63.
12   *Statistics of Military Effort*: 580.

Table 3.1 British Army Rations 1917–18

	Frontline (4,193 calories)	Lines of Communication (3,472 calories)
Bread	16 oz	14 oz
Meat	16 oz	12 oz
Bacon	4oz	3 oz
Vegetables	9 6/7 oz	8 oz
Sugar	3 oz	2 oz
Butter or Margarine	6/7 oz	1 oz
Jam	3 oz	3 oz
Tea	½ oz	½ oz
Cheese	2 oz	2 oz
Condensed Milk	1 oz	1 oz
Rice	–	2 oz

Source: Statistics of Military Effort 1999: 586.

The food detailed in the official records carried the proviso that the army reserved the right to make substitutions when operational conditions demanded, and this was a regular occurrence. For example, by 1916 the War Office had acknowledged that the 'meat resources of the world [were]... more restricted' and cheaper alternatives to fresh and tinned beef were sought. Sausages, brawn and rabbit were used, the latter as an 'occasional substitute', although as 5,649,797 rabbit skins were sold by the army during the war, raising a total of £123,192. This volume suggests that rabbit was rather more frequent than 'occasional'.[13] The accounts of both the men and their officers indicate that substitutions were unpopular but regular occurrences and rarely satisfied the men: items such as sardines were not regarded as an acceptable alternative to meat.[14]

In order to provide the required provisions, the army needed to ensure that the military's priority in the food markets was clearly established. The assertion that military need ranked above that of the civil population was not unanimously accepted and in August 1914, a rash of panic buying broke out and food supplies came under pressure as people hoarded goods. Shops were emptied across the country, and the mayor of Preston was compelled to issue a warning to his community: 'we must condemn the selfishness, almost criminal, of those who have been appropriating the commodities of life to the detriment of the community'.[15] Many producers regarded the high prices on the home market as a far more attractive proposition

---

13　*Statistics of Military Effort* 1999: 850, 580.

14　See Dunn 2003: 302.

15　Cartmell 1919: 30.

than the filling of relatively low-priced army contracts. General S.S. Long, the head of the Army Service Corps (ASC) was quick to step in and forcibly clarify priorities. When civilian contractors were uncooperative, he requisitioned their meat supplies and informed the parliamentary financial secretary at the War Office that an act of parliament must be passed to legitimize his actions. The demand was met that same day, as the prospect of hefty compensation claims for the illegal seizure of foodstuffs proved a powerful incentive for the Government.[16]

Supply Processes

The army had established its right of access to food supplies, but although the principle had been accepted, it would take time for the necessary provisioning processes to be established and in the interim men went hungry. The challenge the army faced in housing, clothing and feeding its recruits was enormous, particularly at the beginning of the conflict, when it was unprepared for the influx of new soldiers. In the early summer of 1914, the regular British Army stood at 247,432 officers and men, by the end of the sixth week of the war, a further 478,893 men had joined up.[17] Whilst training could be deferred, uniforms improvised and accommodation commandeered or rapidly created under canvas, food was a greater challenge: supplies were limited, logistics undeveloped and demand unrelenting. Frederick Gale, who joined The Essex Regiment on 10 September 1914, kept a diary for the first two months of his service, in which he recorded in great detail the food he ate. Gale was a grocer by trade and so may have had something of a professional interest in the rations provided. The situation in the training camp was especially bad in his first days there and on one occasion the meat for dinner ran out, leaving those last in line to make do with potatoes and gravy.[18] However, Gale noted a gradual improvement, and three weeks later was able to write, 'We had a very nice dinner today. We had meat, potatoes and kidney beans. It was quite a luxury to get two vegetables.'[19]

The situation in Flanders at this time was worse; the supply lines for the British Expeditionary Force had completely broken down during the Retreat from Mons in August 1914. Soldiers were forced to rely on their own scrounging skills or the generosity of the local population in order to eat. The war soon settled into its static state, which facilitated the supply of the troops unlike the rapid movement associated with attack or retreat. The trenches of the Western Front were close to the coast where the supply ships landed, but the tonnage of food required was vast, for example, the ration strength in March 1918 stood at 1,828,098.[20]

---

16 Young 2000: 46.
17 Holmes 2005: 103, Simkins 1988: 75.
18 Gale 1914: 10.9.1914.
19 Gale 1914: 1.10.1914.
20 *Statistics of Military Effort* 1999: 756.

Military strategists are clear on the importance of proper provisioning procedures; Basil Liddell Hart, himself a veteran of the conflict, noted that 'strategy depends on supply – without security of food the most dazzling manoeuvres may come to naught.'[21] 'Security of food' embraced a number of different and possibly conflicting pressures. The soldiers' dietary wants had to be balanced with the need to supply other materiel and food requirements had to be considered in the context of not only availability and cost, but also its ease of storage and transport.

The soldiers' food was not the army's highest priority. That position was taken by arms and ammunition without which the army was impotent; men might function to some degree when empty, the guns would not. The men's rations were denied even second position on the priority list by forage for the vast numbers of horses and mules, which formed a critical part of the supply chain by pulling the ordnance wagons.[22] The availability of mechanical transport was limited and although the use of vans and lorries increased, animals remained central to the distribution systems. By late 1918, 382,266 horses and mules were in use on the Western Front and feeding them was a challenge.[23] Unlike petrol for motor vehicles, the animals needed to eat whether or not they were actually being used and each consumed around 25 lbs of food a day.[24] The daily ration figures for the supply of a division show that 120,000 lbs of oats, hay and bran were required for the animals, a tonnage that considerably exceeded that of the soldiers' food.[25] During the war, 5,438,602 tons of forage was shipped from Britain to the Western Front, a figure which slightly exceeded the weight of ammunition transported, but vastly outstripped it in volume.[26] By 1916, the pressures on shipping to transport the quantities required had become unsustainable and efforts were made to grow hay on the abandoned agricultural land near the front or purchase fodder from local farmers.[27]

Nevertheless, the bulk of the food eaten by both animals and men was shipped to the continent rather than purchased locally. Base Supply Depots (BSDs) were constructed near the Channel Ports to store the unloaded cargo. The quantities of food and other essentials held were huge: in one hangar at Le Havre, 80,000 tons of 'articles of supply' were stored at any one time, in a building more than half a mile long and over six hundred foot wide.[28] The scale of the project was vast: in 1918, the Boulogne BSD's highest monthly issue of frozen meat was 21,658,847 lbs.[29] The use of frozen meat by the military was a recent innovation. The army's *1899*

21　Liddell Hart 1999: 75.
22　Thompson 1991: 41.
23　*Statistics of Military Effort* 1999: 878.
24　Beadon 1931: 96.
25　*Supplies*: 25–6.
26　Keegan, Holmes, with Gau 1985: 235.
27　Gibson 2003: 179.
28　Beadon 1931: 98.
29　Young 2000: 390.

Supply Handbook had rejected its use, stating that the freezing process diminished the food's nutritional value.[30] This disapproval had been overcome and supplies of frozen Argentine beef, as well as those from the United States, Canada, Brazil and South Africa, were used from 1914 onwards. Originally the supplies were shipped to Great Britain and inspected there before onward transport to France, but the extra time that this double-handling required became unsupportable and it was decided to ship the meat directly from source to the BSDs.[31] Fresh or frozen meat was often substituted with tinned meat, predominantly bully beef, the majority of which was purchased in South America. Other tinned meat products included 'Pork and Beans', mainly purchased in North America, and tins of 'Meat and Vegetables', which were produced at home.

Problems of Supply

Logistical problems influenced provisioning: food that occupied the least space and had the longest shelf-life was best, at least in terms of supply. T.E. Lawrence wrote that 'the invention of bully beef has modified land-war more profoundly than the invention of gunpowder', the logic being that range was more important to strategy than force and the easily transportable tinned meat, packed with energy values, facilitated rapid troop movement.[32] Bully beef may have excited the strategists, perhaps because those planning troop movements and battle orders at H.Q. rarely ate it, but it had the opposite effect on the men at the front. Much of the dissatisfaction with rations was centred on the unpopular, but ubiquitous, bully. The limits of contemporary nutritional science militated against the ranker. The army's concern was to deliver the requisite energy values to the troops and all too often, the easiest solution to rationing problems was bully beef and biscuit. The high fat content of the former delivered a large number of calories in a relatively small tin. Biscuit, unappealing as it was, had similar energy values and a much longer life than the more popular bread.[33]

Despite the confident tone of the army's postwar assessment, it was aware of shortfalls during the war in the rations provided from the BSDs, in particular in the provision of vegetables. Indeed vegetables implied a variety that was generally absent and in most instances the word was synonymous with potatoes. In the detailed breakdown of supplies sent to a division, the only vegetables specified were 7,500 lbs of potatoes daily and 2,500 lbs of onions four times a week.[34] There is an annotation to the entry which states that if fresh supplies were not available 'dried vegetables or dried fruits or vegetables locally purchased are issued in lieu'.

---

30   *Supply Handbook* 1899: 28.
31   *Statistics of Military Effort*: 842.
32   Shaw 1938: 10.
33   Biscuit is a hard-baked mixture of flour, salt and water.
34   *Supplies*: 25–6.

It was overly optimistic to assume that either fresh local supplies or reserves of the dried items would be available on a sufficient scale. One ranker noted that during the winter of 1916–17, even the great mainstay, potatoes, had been unavailable for months as a poor harvest had resulted in a shortfall also felt on the home front.[35] Subsequently, in the spring of 1917, a series of vegetable gardens in the back areas had been instituted by the army in an attempt to grow its own produce.[36] The official records stated that, by the end of the war, the army in France was 'practically self-supporting in vegetables'.[37] Although as the soldiers appeared to have been offered very few vegetables, the army's proclaimed self-sufficiency may not have been difficult to achieve.

The effort required to work the gardens was not always forthcoming. The gardens were sited at rest and base camps, and the requirement for labour came at a point in his service when a soldier might have hoped to be at leisure. In addition, the cyclical nature of life on the Western Front meant that soldiers had no guarantee that they would be able to reap what they had sown or weeded. In their camp near Elverdinghe, the men of the 10th Essex were exhorted to dig over two acres of land, in order to grow vegetables, but were reluctant to expend energy for the benefit of the other soldiers who would be camped there at harvest time. This account also records that, even worse, such efforts could benefit the enemy, as they had in the Somme area when German troops had overrun British territory and inherited the carefully tended gardens.[38] If cultivated vegetables were unavailable, an imaginative approach to wild sources in the hedgerows was occasionally employed. Eric Hiscock recalled a stay in a convalescent camp in France, where the medical officer concerned by the lack of vegetables in the men's diet, ordered that the nettles surrounding the area should be picked and cooked. The unenthusiastic fatigue parties charged with the task were indiscriminate in their harvesting, picking any green plant they found. It did not take long for sickness to break out and soon 'the whole camp was running to the latrines'.[39]

Army Cooking

Despite the shortages created by problems of supply, the army's publications indicate awareness, at least in theory, of the need to provide its men with variety in their meals. Boredom with the same food leads to its rejection and the possibility of under-consumption and, consequently, under-performance. 'Menu-fatigue' as it is now termed, is a problem with which the British Army continues to struggle, as

---

35 MacDonagh 1935: 15.4.1917; he wrote that he had not eaten a potato 'for weeks'.

36 Smith 1922: 226, 281.

37 *Statistics of Military Effort* 1999: 583.

38 Banks and Chell 1924: 246.

39 Hiscock 1976: 106.

the difficulty of providing appetizing and varied food in war zones over lengthy periods of time remains unresolved.[40] In 1915, a *Manual of Military Cooking and Dietary Mobilization* was published to assist cooks.[41] It contains recipes and a complete, one hundred day menu schedule for use in the training and reserve camps. The main aim of the schedule was not so much to provide a pleasing variety for the men as to ensure that the cookhouses made the most economical use of ingredients. Following the menu plan meant that the leftovers from one day could be included in the next day's dishes. The recipes in the publication sound relatively enticing. On closer examination, they prove, in the tradition of army food, to be variations on a single theme and the theme was meat, whether boiled, fried, stewed, braised, steamed or roasted. Even the optimistically named, 'Fish Paste', was four tins of sardines mixed with eight tins of bully beef.

The provision of meals when the troops were in camp, and permanent cookhouses could be established, was relatively straightforward. Feeding men who were on the move or situated in temporary camps or exposed trenches was a far greater challenge. Field kitchens could be set up and the manuals had complex diagrams illustrating the improvized construction of ovens from material that might be available, such as bricks and corrugated iron. The life of the soldier on the Western Front was one of perpetual movement as they moved between the frontline, the reserve lines and the back areas. Travelling cookers were developed and these could be pulled along on the lengthy marches, ensuring that the men received a hot meal when they stopped to rest. Feeding soldiers in the front areas was particularly difficult given the problems of access through narrow trenches and the danger of smoke from cookers attracting enemy attention. Pan-packs were devised to carry hot food to the men in the frontline. These containers were packed with tins of hot stew and insulating hay and strapped to ration parties' backs, allowing them to keep both hands free to navigate the often treacherous routes forward.

Despite their efforts, subsequent publications indicate that the army was aware all was not well with the provisioning of those men distant from the established cookhouses. In November 1917, a booklet to address the problem entitled *Cooking in the Field* was issued.[42] It described in detail the construction of alternative ovens less demanding of fuel and building materials, such as hay box cookers. In addition there were a number of new recipes, although 'new' was not strictly accurate, as the names of the dishes may have been novel but the ingredients must have induced a sense of déjà vu. Bully beef was the chief constituent, whether in 'Bread Soup' (bully with bread and stock), 'Spring Soup' (bully with vegetables and stock) or 'Potted Meat' (minced bully with pepper).[43] The role of the cooks

---

40   See discussion on current Ministry of Defence website in Messer 2005.

41   *Manual of Military Cooking and Dietary Mobilization* 1915 and reprinted in 1917 and 1918.

42   *Cooking in the Field* S.S.615 1917.

43   *Cooking in the Field* 1917: 17–20.

in alleviating the monotony of the rations was addressed and they were urged to demonstrate initiative in order to provide as varied a diet as possible. The men's well-being was not the sole objective and the preface explained that it was essential that cooks were made aware that it was possible to achieve great savings without in any way stinting the meals.[44] There were reminders on the importance of careful management of supplies and copies of balance sheets were provided in which the smallest amounts of rations issued should be recorded. The significance of fats, in particular, is evident with the introduction of the 'Dripping Account': fats were an expensive foodstuff and cooks were required to keep close control of them.[45] The account included the suet from butchered animals, which could be rendered down into dripping, and the grease they were expected to skim off the top of cold washing-up water, from which glycerine could be extracted for use in the manufacture of explosives.[46]

Difficulties in Consumption and Digestion

Whatever the exhortations to the cooks and the establishment of secure lines of supply, the soldiers' accounts indicate that all too often they were given tinned bully beef and hardtack biscuit, rather than a freshly prepared meal. The army's duty to provide a certain number of calories may have been discharged, but the men were left with a diet that was not only unappealing but had physiological problems associated with its consumption. Biscuit may have been nutritionally similar to bread, but eating it was a very different experience: its rock-like consistency meant that even men with good teeth found it a challenge and for the many men with bad teeth, it could verge on the torturous. The lack of calcium in working-class children's diets and an absence of dental care had resulted in a generation of men whose teeth were not sound enough to deal with the density of army biscuit. The lack of ascorbic acid in the army diet was likely to have exacerbated the problem, although this was not acknowledged at the time. Nutritional science now recognizes that ascorbic acid deficiency falls into several stages the most extreme of which is scurvy. It was only this latter state that was understood and treated by the medical officers of the First World War, as for example in the outbreak amongst the Indian troops in Mesopotamia in 1915.[47] However, in its earlier stage, lack of ascorbic acid results in small haemorrhages, particularly in the gums, infection frequently sets in, causing the teeth to loosen and the mouth to become extremely sore.[48] In a diet where vegetables other than potatoes were rare, and often even those were lacking in the latter years of the war, and fruit, other than jam, was

---

44   *Cooking in the Field* 1917: 1.
45   *Cooking in the Field* 1917: 32–3.
46   *Cooking in the Field* 1917: 34–5.
47   Harrison 1996: 477.
48   Eckstein 1980: 193.

virtually unheard of, it is likely many men were suffering with this intermediate stage of the disease. Additionally, ascorbic acid deficiency impairs wound healing, which must have had serious implications for the boils that seemed to fester interminably and the flesh wounds that obstinately refused to scar, even in the absence of gangrene.

In addition, the bully beef, with its high fat content, presented its own health issues. Fat is not an easily digested food and the presence of high levels of dietary fat cause the secretion of a hormone which inhibits gastric emptying, in order to allow the prolonged time required for its absorption.[49] The soldiers' high fat intake resulted in a slowing of peristalsis, making constipation common place, as the men's numerous references to the dreaded 'No. 9 Pill' indicate.[50] Conversely, diarrhoea was also common in the trenches, generally as a result of dysentery or gastroenteritis, although fear must also have been a contributory factor. As well as the pain and discomfort, the increase in the rate of peristalsis results in the malabsorption of nutrients as the food passes too quickly through the gut. This was especially deleterious when the soldiers' diet was generally rather short of the necessary vitamins and trace elements. The lack of pure drinking water was an additional factor in the prevalence of diarrhoea. The dry and fatty ration relied upon a good supply of water with which to wash it down, something lacking in the trenches. Water is heavy and the transport of large quantities was difficult. Suitable containers were in short supply and inadequately rinsed petrol cans were often used, giving the water an unpalatable taint. It was also common to treat drinking water with chloride of lime, a bleaching substance which may have purified the water but also made it undrinkable. Consequently, men at the front were often thirsty and it was common for soldiers to drink from the puddles in the shell-holes around the trenches. As these could also contain body parts or corpses, the gastric consequences proved serious.

Conclusion

This chapter has explored the practical aspects of army provisioning, and in terms of energy values, the rations were usually sufficient. There were occasions when the supply chain failed completely, but these were rare and generally food was available, although all too often it was bully and biscuit and not the varied diet promised in the official ration scales. The widely held opinion amongst historians of the war that the extra military calories automatically equated to satisfied soldiers depends upon the assumption that eating satisfaction is based solely upon the energy values consumed. Psychologists, anthropologists, sociologists and indeed the soldiers' own accounts tell a far more complex story: food's physiological role is inextricably linked with the overwhelming complexity of its social and

---

49   Eckstein 1980: 53.
50   This was a strong laxative popular with M.O.s.

emotional associations. The army's achievements given the scale of operations were significant, but it is only part of the provisioning narrative. For the civilian soldiers, the rations were distressingly different from their pre-war eating and an inescapable reminder of the military existence forced upon them. Even if the calories had been delivered in a palatable form, it is unlikely that they would have fully compensated the men for the freedom they had sacrificed.

References

Arthur, M. 2001. *When this Bloody War is Over: Soldiers' Songs of the First World War*. London: Piatkus.

Banks, Lieutenant-Colonel T. M., and Chell Captain R.A. 1924. *With the 10ᵗʰ Essex in France*. London: Gay and Hancock.

Beadon, Colonel R.H. 1931. *The Royal Army Service Corps: A History of Transport and Supply in the British Army*. Cambridge: Cambridge University Press.

Burnett, J. 1979. *Plenty and Want: A Social History of Food in England from 1815 to the Present Day*. London: Scolar Press.

Cartmell, H. 1919. *For Remembrance: An Account of some Fateful Years*. Preston, G. Toulmin & Sons.

Cole, H.N. 1984. *The Story of the Army Catering Corps and its Predecessors*. London: privately published.

Cooking in the Field SS 615. 1917. London: HMSO.

Drummond, J.C., and Wilbrahim, A. 1958. *The Englishman's Food: A History of Five Centuries of English Diet*. London: Cape.

Duffett, R. 2008. A War Unimagined: Food and the rank and file soldier of the First World War, in *British Popular Culture and the First World War*, edited by J. Meyer. Leiden: Brill.

Duffett, R. 2011. *The Stomach for Fighting: Food and the Soldiers of The Great War*. Manchester: Manchester University Press.

Dunn, Captain J.C. 2003. *The War the Infantry Knew 1914–1919*. London: Abacus.

Dwork, D. 1987. *War is Good for Babies and Other Young Children: A History of the Infant and Child Welfare Movement in England 1898–1918*. London: Tavistock.

Eckstein, E. 1980. *Food, People and Nutrition*. Westport: AVI Publishing Co.

Floud, R., Wachter, K., and Gregory, A. 2006. *Height, Health and History: Nutritional Status in the United Kingdom, 1750–1980*. Cambridge: Cambridge University Press.

Gale, F. 1914. Diary in the Essex Regimental Archive, Chelmsford.

Gibson, C. 2003. The British Army, French farmers and the War on the Western Front 1914–1918. *Past and Present*, 180, 175–239.

Gregory, A. 2008. *The Last Great War. British Society and the First World War*. Cambridge: Cambridge University Press.

Harrison, M. 1996. The fight against disease in the Mesopotamia Campaign, in *Facing Armageddon: The First World War Experience*, edited by H. Cecil and P. Liddle. London: Leo Cooper.

Hiscock, E. 1976. *The Bells of Hell Go Ting-A-Ling-A-Ling*. Nairn: Arlington Books.

Holmes, R. 2005. *Tommy: The British Soldier on the Western Front 1914–1918*. London: Harper Perennial.

Keegan, J. 2004. *The Face of Battle*. London: Pimlico.

Keegan, J., Holmes, R. and Gau, J. 1985. *Soldiers. A History of Men in Battle*. London: Guild Publishing.

Liddell Hart, B.H. 1999. *Thoughts on War*. Staplehurst: Spellmount.

MacDonagh, M. 1935. *In London during the Great War: The Diary of a Journalist*. London: Eyre and Spottiswoode.

Manual of Cooking and Dietary Mobilization. 1918. London: HMSO.

Messer, P., Owen, G. and Casey, A. 2005. *The Commander's Guide to Nutrition*, http://www.mod.uk/DefenceInternet/MicroSite/DES/OurPublications/CateringPublications/TheCommandersGuideToNutrition.htm.

Roberts, R. 1971. *The Classic Slum: Salford Life in the First Quarter of the Century*. London: Penguin.

Shaw, Lieutenant-Colonel, G.C. 1938. *Supply in Modern War*. London: Faber & Faber.

Simkins, P. 1988. *Kitchener's Army: The Raising of the New Armies 1914–16*. Manchester: Manchester University Press.

Skelley, A.R. 1977. *The Victorian Army at Home. The Recruitment and Terms and Conditions of the British Regular, 1859–1899*. London: Croom Helm.

Smith, A. 1922. *Four Years on the Western Front*. London: Odhams.

Smurthwaite, D. 1993. A Recipe for Discontent, in *The Victorian Soldier: Studies in the History of the British Army 1816–1914*, edited by M. Harding. London: National Army Museum.

Statistics of the Military Effort of the British Empire during the First World War. 1999. London: Naval and Military Press.

Supplies and Supply Transport in the 38th (Welsh Division) by a Senior Supply Officer. Undated. London, publisher unknown.

Supply Handbook for the Army Service Corps. 1899. London: HMSO.

Thompson, Major General J. 1991. *The Lifeblood of War. Logistics in Armed Conflict*. Oxford: Brassey's.

Young, M. 2000. *Army Service Corps 1902–1918*. Barnsley: Leo Cooper.

Chapter 4

Fighting a Kosher War: German Jews and Kashrut in the First World War

Steven Schouten

Introduction

This chapter deals with the impact of the First World War on the kosher foodways of German Jews.[1] Were Jews able to uphold, foster and enact their dietary laws in wartime Germany? How did they organize themselves in this respect? And to what degree was there a change from pre-war times?

In providing answers to these questions, I will first consider the significance of the dietary laws, or the laws of kashrut, to Jewish food practices in prewar Germany. I will then concentrate on kashrut in wartime Germany on both the home and battle fronts, particularly the latter. In doing so, I will explore the role of gender in kosher foodways in a war that was characterized by a division between a female home and a male battle front. Ruth A. Abusch-Magder has pointed to the centrality of Jewish women in the day-to-day enactment and definition of kashrut in Imperial Germany before 1914.[2] This chapter uses her findings, but also addresses the role of men in the daily enactment of kashrut in wartime Germany. Although women remained crucial to the continuance of 'homemade Judaism', the war also seems to have strengthened male authority and influence in decisions related to the ultimate control and enactment of kashrut. Men (re)gained power at a central level through the work of rabbis and Jewish (Orthodox) organizations, but they also *actively* influenced and shaped the practice of the enactment and definition of kashrut at the battlefront. Unwilling to relinquish their Jewish identities, Jewish soldiers, especially the traditional and Orthodox, not only appreciated and fostered kosher foodways, but they also sought ways to express and implement them (semi-)independently from, and simultaneously to, their wives and mothers at home.

---

1   Due to limited space, this chapter will concentrate on German Jews and not on the broader group of Jews living in Germany, hence there will be only limited exploration of the foodways of East European Jews, many of whom had settled in Germany since the 1880s.

2   Abusch-Magder 2005: 169–92, Abusch-Magder 2006a.

Kashrut in Imperial Germany: 1870–1914

On the eve of the First World War, Jews constituted no more than one per cent of the total population in Imperial Germany (615,000 of 65 million people in 1910). The majority lived in the rapidly growing and expanding urban centres: Berlin, Hamburg and the Ruhr area. Although acculturated into the values, norms and lifestyles of their Gentile environments, most Jews characterized themselves by distinct social-cultural practices, one of which included the enactment of kashrut.

Regulations on eating are found in a number of texts from the Torah. In Judaism, dietary laws are comprised of three categories of regulations: those defining the animals which may or may not be consumed (Lev 11 and Dt 14); those prohibiting the consumption of blood (Lev 17, Gen 9:4–6); and those prohibiting the cooking of a calf in its mother's milk (Ex 23:19, 34:26, Dt 14:21). The degree of the enactment and the interpretation of these laws in society varied across time and by geography, but they were generally not very strictly applied by the majority of Jews in Imperial Germany.[3]

Due to their Emancipation in the early nineteenth century, most Jews in this country had distanced themselves from their history and traditions. Aware that 'otherness' in German society obstructed integration, they had adapted their foodways and lifestyles to the cultural demands and codes of an emerging industrialized, bourgeois society. The loss of social control following the steady Jewish migration to urban centres challenged rabbinical authority and stimulated a sense of distance from, and the consequent abandonment of, halachic law and strict kashrut. The rise of a modern and state controlled food industry further complicated kashrut. As a result, the enactment of the dietary laws increasingly became a matter of choice for individual Jews. Some decided to keep kosher, others no longer held back from eating pork or from adding elements of pork-based fats to their meals. A significant minority of traditional and Orthodox Jews, however, remained deeply attached to the enactment and strict interpretation of the dietary laws. Kashrut increasingly served as a divisive issue to separate these Jews from the reform-minded, Liberal and more assimilationist Jews.

In keeping with the decline of halachic law and rabbinical authority was, as Abusch-Magder writes, a relative increase in female power regarding the choices related to the enactment, degree and definition of kashrut, although still within the framework of the limitations imposed by a predominantly patriarchal society. Responsible for the home-based tasks, mothers and wives decided what foods to select, prepare and serve for their Jewish families.[4] According to Marion A. Kaplan, many women, especially in the urban centres of Imperial Germany, regarded kashrut as a 'time-consuming extra effort and eagerly relieved themselves from its burdens.'[5] In rural areas, however, social control and ties to the parental religious foodways often remained strong, even though, here too, distance from a traditional

---

3   Kraemer 2007.
4   Abusch-Magder 2005: 169–92, esp. 169–70, Abusch-Magder 2006a.
5   Kaplan 1991: 71.

and halachic past was noted.[6] In the large towns, moreover, Orthodox women unconditionally stuck to the strict enactment of the dietary laws.[7] Responding to the overall decline of traditional Judaism and to the ascent of Liberal, reform-minded Judaism, these women aimed to connect acculturated eating patterns with a revitalization of religious and halachic foodways.

Jews in Imperial Germany, and not only the traditional and Orthodox, remained tied to 'kosher' foodways in an ethnic sense as well. Understanding Jewish eating in terms of what Joelle Bahlout has called 'ethnic emblems', Marion A. Kaplan points to the existence of a distinct, "Jewish" eating culture in Imperial Germany beyond a traditionally religious one.[8] Indeed, although many Jews became immersed in German culture and distanced themselves from halachic law, they still clung to the celebration of the Sabbath and other holidays as well as to the traditional Jewish foods and dishes connected with those traditions, like haroset (the Passover sauce) or cholent (*Schalet*, the Sabbath stew), even though the meat slaughtered for that stew was no longer necessarily kosher. Assimilated Jews also continued to use chicken fat and goose fat (*schmaltz*), furthering a tradition born out of religious dietary laws in a secular, ethnic setting.[9] They also retained specific traditions of (milk-free) bread baking during Jewish holidays and festivities, like that of the production of the oval braided *Barches* (Challah) for the Sabbath and that of matzos at Passover. Acculturated urban Jewish girls, moreover, were expected to acquaint themselves with traditional Jewish cooking before they married as a means to maintain a specifically ethnic and religious identity, even while many of these women no longer defined their 'Jewishness' in their ancestor's terms.[10]

Accordingly, on the eve of the First World War, kashrut was in decline, yet it continued to play a central role for many German Jews and it manifested itself in a variety of religious and ethnic ways.

Home Front: 1914–18

Germany faced similar problems of food production and supply during the First World War as other European countries. However, a particular characteristic of the food situation in Germany since the end of the nineteenth century was the strong dependency of city dwellers on animal products and imported foods. Industrial and technological advances had resulted in geographical distance between producers and consumers which led to huge problems when the infrastructure broke down during the war. Fuel for domestic transport was in severely short supply,

---

6 Seligsohn. 1903, Part II, 84 (Leo Baeck Institute, New York, ME 596).

7 Straus 1962.

8 Kaplan 1991: 72.

9 See, for example: Letter by Betty Scholem [21.IX.1917] in Scholem and Scholem 1989: 16.

10 Kaplan 1991: 72–4, Abusch-Magder 2006a, Abusch-Magder 2006b: 263.

particularly after 1916, and this complicated food deliveries. In addition, quantities of food were lost during storage or transport. Due to Germany's dependency on imported foods, Great Britain launched a war of economics, and through its blockade of German ports, seriously depleted its food supply. Military tactics were increasingly influenced by the immediate need for food, and Germany's submarine war campaign of 1916–17 aimed to run the British blockade in order to make way for desperately needed foodstuffs. Germany's strategy failed, however, and from mid-1917 through to the end of the war, it depended almost entirely on its limited home-produced food supplies. On the eve of the war, Germany imported about one third of its overall food supply, including important staples of the pre-war German diet, such as milk, butter and meat, as well as coffee. Germany was strongest in supplying its own carbohydrates through potato and rye cultivation, but due to an early frost, the 1914 harvest yielded a poor supply of the potatoes central to the German diet. Farmers and merchants raised prices and held back supplies, exacerbating the problems. In addition, bad winters over the course of war wrought further havoc in domestic potato harvests, resulting in turnips being substituted for potatoes in the winter of 1916–17, a variation in diet that hit the bulk of the population very hard.[11]

Scarcity and high prices created problems for all, but especially for the urban working class. The price of potatoes had nearly doubled by October 1914 and meat prices began to rise in November 1914, as suppliers could sell it at extremely high rates to the army. Pork, dominant in the German diet, had become unaffordable to the urban poor by February 1915. All this had crucial implications for the people's health: 'If Germans did not starve in the war,' Belinda J. Davis writes, 'certainly the health of the poorest urban dwellers was seriously compromised as a result of the nutritional deficiencies of the war era and in combination with other conditions and deprivations.'

Although the government tried to resolve these matters, it failed in its aims, despite fixed prices, together with the introduction of ration coupons for bread, soon becoming the norm for all foods.[12] Bread cards and the substitution of normal bread for corn bread were part of wartime foodways by early December 1914.[13] Pressed by various groups in society, the government adopted a policy of total food control in deference to the needs of those of the population 'of lesser means'.[14] From the spring of 1916 the first soup kitchens and meal halls (public dining) for the working and lower-middle class were created, housed in schools and run by charitable organizations. Concentrated in working-class districts, they targeted working-class women. The state also hoped to attract a larger clientele in due course, but this failed to materialize.[15]

---

11 Davis 2000: 50, Geyer 1998: 47, Albrecht 1968: 26–9, Bouzon and Davis 1997.
12 Davis 2000: 22, 48–70, esp. 48–50, 69.
13 Klemperer 1996, 276–8.
14 Davis 2000: 93–5.
15 Ibid.: 144–5.

The food crisis complicated the ability to keep kosher, especially in the large towns. Although not impossible during the war, kashrut in urban life was difficult because of food scarcity and high prices; kosher meat, for example, became expensive and rapidly unaffordable for lower-middle and working-class Jews. It became a lucrative business for the uncontrolled black market, where the lack of provenance also increased insecurity about the kosher status of meat.

Yet various traditional and Orthodox Jews, especially amongst the middle or upper-middle class, found ways to keep kosher during the war. Gershom Scholem (1897–1982), after being thrown out of his father's house, spent time in 1917 in Pension Struck in the Uhlandstrasse in Berlin, where its hostess ran, in Scholem's words, 'a strict kosher household'.[16] Rahel Straus (1880–1963), a bourgeois middle-class Orthodox Jewess from Munich, records that she maintained a kosher home during the whole war. She had produce from her little garden in the countryside, however, and also benefited from contacts with peasants and rural acquaintances.[17]

Urban lower-class Jews encountered more difficulties with regard to the enactment of kashrut than the middle and upper-middle class. East European Jews [*Ostjuden*], large numbers of whom had migrated during the war to Germany to work in the war industry, constituted a substantial part of this group. Between 1916 and 1920 70,000 Jews from Russia and Poland moved to Germany, two thirds of whom were industrial workers.[18] Jews from Eastern Europe had already settled in large numbers in Germany in the decades before the war, but now non-German citizens constituted about a quarter of Berlin's Jewish community and they were an actual majority in such communities as Dresden, Leipzig and Hamburg. Halachic authorities were shocked by the limited degree of ritual observance by these working-class Jews, although some of these authorities also noted that some workers refused to eat *treyf* [non kosher] foods in the canteens at work.[19] Immigrants with families, Gertrude Weil-Welkanoz notes, were often compelled to send their children to soup kitchens, some of which offered kosher foods.[20]

In defense of a kosher lifestyle various Jewish organizations and citizens came to the aid of fellow Jews from the very beginning of the war. Jewish women's groups cooperated on the local and national level with other women's groups and government agencies, and they were active in the distribution of kosher food at home and at the front. In the East, the Jewish National Women's Service donated funds to organize kosher kitchens in Posen.[21] Individual Jewish women, too, such

---

16 Scholem 1997: 93.
17 Straus 1962: 209.
18 Maurer 1986: 95, Heid 1995: 12.
19 Heid 1995: 301, 486, 522. On the *Ostjuden* and pre-war immigration, see also Wertheimer 1987.
20 Weil 1930: 282.
21 Kaplan 1991: 221–2.

as Margarethe Goldstein (1885–1960), endorsed the ideals of 'social mothering' by organizing soup kitchens and sending food packages to men at the front.[22]

With regard to these aims, Mordechai Breuer has pointed to the central role played by the *Freie Vereinigung für die Interessen der Orthodoxe Juden* [Free Association for the Interests of the Orthodox Jews]. Although a variety of Jewish organizations did similar work, for example the Liberal *Verband der Deutschen Juden* [Association of German Jews] and the rabbinical organizations, the *Freie Vereinigung* was the most successful in terms of both efforts and results because of its diplomacy and contacts with the government. The association, which could count on the respect of many Liberal Jews as well, founded, amongst others, the *Kriegskommision für rituelle Lebensmittel* [War Commission for Ritual Foods], responsible for the preparation of kosher fats and other rationed foods.[23] It also tried to convince the companies responsible for war production to include kosher foods in their canteens and to create separate dining spaces for Jews on the Sabbath or religious holidays.[24] The association's work to benefit the *Ostjuden* at home was secondary, however, to its desire to preserve the religious lives of German Jews at the front by helping them to conform to ritual observance.

Battlefront: 1914–18

Food problems also existed on the battlefront. Generally, soldiers enjoyed better food, particularly at the beginning of the war, than those at home due to the army's tight supervision of food provision for the soldiers and the state's policy of serving the soldiers first. Victor Klemperer (1881–1960) enlisted at the front in July 1915 when 'the food [was] still very good and abundant' and soldiers ate lots of 'sausage, cheese and chocolate', the latter consumed in large quantities and numerous varieties.[25] 'Bread, cheese and butter' were 'generously distributed' and their distribution was, according to Klemperer, 'well-organized'.[26] If soldiers complained about food it was not so much about its availability, but rather about its unequal distribution which often enabled 'respectable people' (e.g. officers), to receive 'an extra sausage'.[27]

Other soldiers, however, emphasize the limited quantity and moderate quality of their foods, and army bread was criticized by many. Some men also stressed the need to buy additional foods from the local population, a matter complicated

---

22   Ibid.: 222.
23   Breuer 1986: 344.
24   Heid 1995: 306.
25   Klemperer 1996, 96: 294 and 299.
26   Ibid.: 335.
27   Ibid.: 348. On the association of officers and better food, also 373, Toller 1978: 65, Zuckmayer 2002: 260–61.

from the start of the war by exorbitant prices.[28] Civilians, as mentioned above, often sent soldiers food packages, which frequently included essential rather than additional foods.[29] Many soldiers actually requested that their family or friends at home send them extra food or money.[30] It was soldiers from rural backgrounds who tended to receive the largest packages.[31] This extra food could both increase tensions with soldiers from urban backgrounds yet also strengthen solidarity. Ernst Simon (1899–1988), a Jewish soldier from Berlin, records how Nathan Kahn, a Jew from a rural background, shared his food with him while they were at the front.[32]

The collapse of the German distribution system and the impact of the blockade hit the food situation hard on the battlefront as well. Soldiers experienced reductions in their food rations and they were conscious, especially by 1917, that their diet had become increasingly plain.[33] They continued to receive foods from the army and their families at home, but also appeared to develop a constant fixation on food and eating. In this context, ongoing differences in foods between soldiers and officers further stimulated the resentment in the ranks.[34]

Food was an immediate preoccupation for all German soldiers, but to a minority group with particular dietary requirements it assumed an extra dimension. Numbering about 1 per cent of the German Army, Jews had little choice but to accept – if not adapt themselves to – *treyf* circumstances.[35] Although they could attend Jewish holiday services and visit rabbis at various locations and times in the war, it was very difficult to uphold kashrut, especially in the frontline and in isolated postings. Meals were centrally prepared by Gentile cooks and with products that had been distributed by the army. Pork based sausages and bacon were central ingredients of the daily diet and were often included, as Carl Zuckmayer writes, in so-called vegetable soups.[36] As cooks in the German army were generally not Jewish,[37] and as the food supply and its distribution was in the hands of the state, Jews did not control the foods that were selected and prepared

---

28 Letter from Leo Jarecki (Jaretski) [1.III.1918], Letter from Karl Levit [16.V.1916], Letter from Julius Marcus [15.VI.1915], in Hank and Simon 2002 I: 303–4, II: 417 and 478.

29 Zuckmayer 2002: 260, Toller 1978: 61.

30 Letter from Leo Jaretski [17.X.1915], ibid. [28.VI.1917], ibid. [1.III.1918], Letter from Benno Jastrow [12.VI.1915], in Hank and Simon 2002 I: 298–9, 304 and 309.

31 Ziemann 2007: 76–7.

32 Letter from Ernst Simon [20.IV.1918] in Simon 1998: 7–9.

33 Zuckmayer 2002: 279, Letter from Karl Levit [23.I.1917] in Hank and Simon 2002 II: 424.

34 Toller 1978: 65.

35 Segall 1922, Berger 2006: 177, 181.

36 Zuckmayer 2002: 257, on the mixing of foods, see also Letter from Siebert Jungmann [4.VIII.1917] in Hank and Simon 2002 I: 342.

37 Cooks were not professionals, but were recruited from the ranks, see Zuckmayer 2002: 257.

for them. Kashrut, or a limited degree of kashrut, was only possible if there were a religious Jewish cook and a substantial group of Jewish soldiers, but this was rarely the case. Moreover, meat served by the German army was, for most of the war, not ritually slaughtered. Unhygienic conditions during cooking and consumption as well as the absence of multiple sets of dishes, pots, bowls, cutlery and pans further restricted the strict observance of kashrut at the front.

Jews were well aware of the limitations war imposed upon their eating habits and food practices. Having lived for centuries among Catholics and Protestants, however, and often away from their kosher homes on business trips, Jewish men had been familiar with survival in *treyf* circumstances. During the first days of the war, many Jews had already anticipated the difficulties of keeping kosher during wartime. Some Jews even believed that it was best to get used to *treyf* food as soon as possible: 'In the field we shall have to eat what is given to us', one German Jewish soldier was told, 'and it is to the benefit of our fighting ability to get used to *treyf* food [in military canteens] as soon as possible.'[38] Intended to also emphasize the patriotic orientation of Jews, these words reveal that acceptance of *treyf* food was seen as a necessity and part of warfare.

This is not to say that Jewish soldiers were unconcerned about kashrut during the war. On the contrary, traditional and Orthodox soldiers, above all, continued to foster kosher foodways and they also presented rabbis with questions about some of the difficulties they encountered in their pursuance of living a kosher lifestyle.[39] Unlike the male East European lower-class workers at home, many of these German traditional and Orthodox soldiers clung to their faith in reaction to the loneliness of war and the war experience.[40]

Despite the practical challenges of combining military service with specific dietary requirements, Jewish soldiers did manage to obtain kosher meals and foods in the First World War. Although strict kashrut was fairly impossible in the frontline, traditional and Orthodox soldiers took advantage of a whole range of possibilities to help them keep kosher throughout the war.

First of all, Jewish soldiers received kosher foods from their families at home and from various Jewish organizations. Often these foods were sent for the Sabbath and other Jewish holidays. One soldier received kosher chocolate from his family shortly before the Sabbath.[41] Another received a *Tatscher* (e.g. *Barches*, Challah; Sabbath bread).[42] A few days before Passover in April 1915, Robert Ziegel received matzos from his family.[43] The Orthodox *Freie Vereinigung* had been allowed by the state to produce canned kosher meat, for distribution at the front, and it had been successful in the granting of an extra financial allowance for religious Jewish soldiers to

38   Letter from 'Baal Milchomo N.' [1914] in Tannenbaum 1915: 11–12.
39   Breuer 1986, 345.
40   Ibid.: 346.
41   Letter from an unknown soldier [IX.1914] in Tannenbaum 1915: 39.
42   Letter from an unknown soldier [14.X.1914] in Tannenbaum 1915: 80, 82.
43   Berger 2006: 150.

enable them to buy kosher foods.[44] Shortly before the Sabbath, the association also distributed Kugel [*Kuchel*] and Apple Schalet [*Apfelschalet*], the latter being a pastry delicacy among Jews from Southern Germany and the Alsace.[45]

Field rabbis, too, of whom some 30 were active during the war, played a central role in the distribution of kosher food.[46] On Jewish holidays they organized kosher meals for Jewish soldiers with the consent of the German army, who had been convinced of their necessity. On such holidays many Jewish soldiers were granted leave to attend services, which were frequently organized in churches, and it was often after such ceremonies that soldiers were able to eat kosher foods.[47] In September 1916, for example, Rabbi Martin Salomonski (1881–1944) organized Jewish New Year meals near St. Quentin in France for no less than 1,600 German Jewish men. Stationed in a local factory, he served 350 men in five groups within 2 ½ hours. Other than on Yom Kippur, all men received hot meals, including soup, vegetables, potatoes and meat from 'reliable hands'. Salomonski managed to do all this with the aid of only three Jewish cooks, seven female volunteers and various (male) soldiers, the latter of whom served the food. Officers and doctors dined at separate tables, but they ate the same food and in so doing, kosher food served as a unifying force between ranks.[48]

On a smaller scale, many Jews also sought ways to eat, at least partly, kosher food at the front, especially on the Sabbath or Jewish feast days. Emil Salomon, a field deputy doctor who had been stationed near St. Juvin, gathered six Jews on 8 December 1914 for Hanukkah, one of whom was a cantor [*Kantor*] and therefore allowed to slaughter a chicken, and another, who was a cook from Paris, then prepared the bird.[49] At another location on the Western Front various Jews handed in all the foods they had received from their families to make a common meal and share a kosher moment on the occasion of Rosh Hashanah.[50] Leo Leßmann from Hamburg records that he and some fellow Jewish soldiers made a 'nice soup' on Rosh Hashanah from canned vegetables and potatoes.[51] Unlike the organized events, these spontaneous meals on a smaller scale were not consumed at tables; soldiers could not clean their cups and they ate Sabbath sardines with their hands and canned meat directly from the tins.[52] Soldiers realized that in 'normal' times

---

44 Breuer 1986: 344.

45 Letter from an unknown soldier in Tannenbaum 1915: 14.

46 Berger 2006: 141 and 158, Letter from an unknown soldier [c.1914] in Plessner 1918: 7–8.

47 Letter from an unknown soldier [1/2.X.1914], letter from Benno Kahn,[undated], letter from Felix Rosenblüth in Berlin [2.IV.1915] in Tannenbaum 1915: 68, 77, 183.

48 Salomonski 1918: 18–20, Salomonski 1917: 61–3.

49 Letter from Emil Salomon [8.XII.1914], in Tannenbaum 1915: 131.

50 Letter from an unknown soldier [29.IX.1914], in Tannenbaum, 1915: 36.

51 Letter from Leo Leßmann [undated], in Tannenbaum 1915: 65.

52 Letter from an unknown soldier [c.1914], in Plessner 1918: 14.

few rabbis would consider their foodways kosher, but they were kosher enough for many of these men.[53]

Jews at the front encountered difficulties in conforming to the requirements of milk-meat separation as well as in the problems of avoiding non-ritually slaughtered or forbidden meats. Abstention from certain prohibited foods, however, was not impossible, as long as these foods were not added to centrally prepared meals or soups. Julius Marx, a Liberal Jew, records a Jewish soldier with a rural background who refused to eat a piece of ham offered by fellow soldiers.[54] Another German Jewish soldier, stationed on the Western Front, rejected the consumption of rabbit, when one day he had wanted to buy bread ['la pain'] from a French peasant, but had been misunderstood and had almost been sold a rabbit ['lapin'] instead.[55] A number of traditional and Orthodox soldiers on the frontline also fasted on the celebration of Rosh Hashanah and Yom Kippur, thus respecting halachic law in another way rather than by the act of kashrut (in terms of the specific food regulations) alone.[56]

In the East, where Germans fought against the Russians, German Jewish soldiers often encountered local East European Jews, many of whom offered these soldiers opportunities to keep kosher. At odds with the anti-Semitism of the Russian state, which had been heightened from the end of the nineteenth century, *Ostjuden* were frequently positive about the more liberal German state and invited German Jewish soldiers to share kosher meals in their homes. Steeped in *tzedakah* or Jewish charity, these invitations were evidence of longstanding Jewish traditions.

On occasion, traditional and Orthodox German soldiers actively looked for East European Jewish families that would share kosher meals with them.[57] Others profited from the presence of the various kosher restaurants in Poland and Russia. For example, Marx, who was mentioned above, paid a visit to a kosher restaurant in Bialystok on 27 August 1915.[58] In the West, similar opportunities for meals at the homes of local Jews were seized by German Jewish soldiers, especially in the Alsace where a considerable number of traditional Jews lived.[59]

Complete enactment of the dietary laws was virtually impossible in the frontlines, as we have seen, but it was possible behind them. Stationed as an assistant doctor on the Eastern Front, the Orthodox Abraham Fraenkel (1891–1965) records that he could not always observe the Sabbath during the war, but he was very proud to have

---

53 Letter from an unknown soldier [14.X.1914] and letter by Hugo Henle [10.XII, 1914], in Tannenbaum, 1915: 80, 136, letter from an unknown soldier [c.1914], in Plessner, 1918: 14.

54 Marx 1939: 52.

55 Letter from an unknown soldier [4.XI.1914], in Tannenbaum 1915: 108.

56 Marx 1939: 84, Letter from 'Fritz' [23.IX.1914], in Tannenbaum 1915: 34.

57 Letter from an unknown soldier [29/30.XI.1914], in Tannenbaum 1915: 120–21.

58 Marx 1939: 78.

59 Letter from Fritz Becker [27.IV.1916], in Hank and Simon 2002, I: 71.

always kept kosher in spite of the difficulties.[60] Fraenkel does not describe how he managed to keep kosher, but stationed far behind the frontline in an area with many traditional Eastern European Jews, he had greater opportunity to choose and organize his own foodways than those afforded to his fellow Jewish soldiers at the front.

Still, there always remained a price limit for soldiers, even when kosher foods were to be found. Herbert Czapski (1896–99), stationed in Warsaw, was unable to buy kosher sausages at the local Jewish meat market in May 1918, as they had become too expensive for him since the German-Russian peace.[61] Leo Jaretski (b.1888) complained about the impossibility of affording *schmaltz*, which had become extortionately expensive in the East by the autumn of 1915.[62]

Jews were attached to kosher foodways, but for other reasons than the religious alone: most importantly, food reminded soldiers of home. The attempts of various soldiers to produce special meals on the Sabbath and on other Jewish festivities not only served as a means to respect Jewish tradition and halachic law, but created an atmosphere of homeliness as well.[63] Such an atmosphere was also established through special Jewish eating places in military canteens, which existed in some German battalions.[64]

Homeliness, moreover, was expressed through the contact with Jews from 'hostile' territory both in the West and in the East. On 18 April 1916, Fritz Becker (b.1895), a Jewish soldier from Berlin born to traditional Jewish parents from Posen, shared a Seder meal with local Jews from a French village and immediately felt as if he were at home.[65] Max Marcus, stationed in the East, records the homely atmosphere and hospitality of the local Jews who had invited him for a meal in the Polish town of Szeczemin in January 1915.[66] Marcus was also invited by a local Jewish family to share a Sabbath meal in Tuschin, near Lodz.[67] Although it is not explicitly stated by the soldiers, the homely character of such meals was possibly strengthened by the excitement of a temporary restoration of traditional pre-war role patterns, where women were responsible for the preparation of meals and men for its consumption.

---

60   Fraenkel 1967: 128.

61   Letter from Herbert Czapski [6.V.1918], in Hank and Simon 2002, I: 182.

62   Letter from Leo Jaretski [17.X.1915], in Hank and Simon 2002, I: 298.

63   Letter from Martin Feist [2.XI.1914], in Reichsbund Jüdischer Frontsoldaten 1935: 19.

64   Letter from Max Sichel [3.XI.1914], in Tannenbaum 1915: 104.

65   Letter from Fritz Becker [27.IV.1916], in Hank and Simon 2002: 71.

66   Letter from Max Marcus [18.I.1915], in Tannenbaum 1915: 163.

67   Letter from Max Marcus [11.I.1915], in Tannenbaum, 1915: 154–5, on the creation of a homely atmosphere by *Ostjuden*, see also: Letter by M.v.d.W. [26.IX.1914], in Tannenbaum 1915: 47–9, on homeliness and Jewish food, see also: Gronemann 1924: 54–5 and 61.

Conclusion

Food, and its shortages, played a crucial role in the lives of Gentiles and Jews in the First World War. For German Jews, as we have seen, it was not only a means of necessity for survival, but also of a way to confront, defend, express and re-negotiate German Jewish identity. I have shown that war sustained pre-war gendered food patterns by further anchoring women into the roles of the gatekeepers of everyday enacted forms of kashrut.[68] I have also shown that the emergence of an (almost) exclusively male battlefront created a situation in which men were forced to play active roles in German Jewish culinary life at the front. As the state controlled both food deliveries and food preparations, Jewish soldiers had to organize themselves to uphold and sustain at least a part of their Jewish identity in wartime Germany. Although in exceptional circumstances, the First World War not only made these men aware of their needs related to kosher foodways, but it also transformed them at various locations into cooks and *active* co-gatekeepers of everyday enacted forms of kashrut. They selected, prepared, and served their own foods, and they also employed a variety of other strategies to fulfill the demands of halachic law and kashrut. These included the rejection of specific foods such as pork and rabbit, self-initiated visits to kosher homes and restaurants in Eastern Europe, and abstention during religious holidays. Despite the fact that the soldiers were aware that their wartime foodways were not always strictly kosher, their actions were nonetheless serious attempts to fight a kosher war. In doing so, they actively joined with their female counterparts on the home front, albeit in a modest way without either the desire or the ability to completely deprive the women from their de facto powers in the everyday choices related to the enactment of kashrut.

References

Abusch-Magder, R.A. 2005. Kashrut: The Possibility and Limits of Women's Domestic Power, in *Food and Judaism*, edited by L.J. Greenspoon, R.A. Simkins and G. Shapiro. Omaha: Creighton University Press, 169–92.

Abusch-Magder, R.A. 2006a. Home-made Judaism: Food and Domestic Jewish Life in Germany and the United States, 1850–1914, Unpublished dissertation, Yale University.

Abusch-Magder, R.A. 2006b. Kulinarische Bildung: Jüdische Kochbücher als Medien der Verbürgerlichung [Culinary Self-Formation: Jewish Cookery Books as a Medium of Embourgeoisement], in *Deutsch-jüdische Geschichte als Geschlechtergeschichte: Studien zum 19. und 20. Jahrhundert* [German-

---

68   There is a wide literature on the use of the gatekeeper-concept, which need not be repeated here. A critical and insightful approach to that literature can be found in McIntosh and Zey 1998: 125–44.

Jewish History as Gender History: 19th and 20th Century Studies], edited by K. Heinsohn and S. Schüler-Springorum. Göttingen: Wallstein Verlag, 159–76.

Albrecht, W. 1968. *Landtag und Regierung in Bayern am Vorabend der Revolution von 1918* [Diet and Government in Bavaria on the Eve of the 1918 Revolution]. Berlin: Duncker & Humblot.

Berger, M. 2006. *Eisernes Kreuz und Davidstern: Die Geschichte jüdischer Soldaten in deutschen Armeen* [The Iron Cross and the Star of David: the History of Jewish Soldiers in German Armies]. Berlin: Trafo.

Bonzon T. and Davis, B. 1999. Feeding the Cities, in *Capital Cities at war: Paris, London, Berlin, 1914–1919*, edited by J. Winter and J.-L. Robert. Cambridge: Cambridge University Press, 305–41.

Breuer, M. 1986. *Jüdische Orthodoxie in Deutschland, 1871–1918. Sozialgeschichte einer religiösen Minderheit* [Jewish Orthodoxy in Germany, 1871–1918. The Social History of a Religious Minority]. Frankfurt a/Main: Jüdischer Verlag bei Athenäum.

Davis, B.J. 2000. *Home Fires Burning: Food, Politics, and Everyday Life in World War I Berlin.* Chapel Hill: University of North Carolina Press.

Fraenkel, A. 1967. *Lebenskreise: Aus den Erinnerungen eines jüdischen Mathematikers* [The Circles of Life: the Memories of a Jewish Mathematician]. Stuttgart: Deutsche, Verlags-Anstalt.

Geyer, M.H. 1998. *Verkehrte Welt: Revolution, Inflation, und Moderne: München, 1914–1924* [An Inverted World: Revolution, Inflation and Modernity: Munich, 1914–1924]. Göttingen: Vandenhoeck & Ruprecht.

Gronemann, S. 1924. *Hawdoloh und Zapfenstreich: Erinnerungen an die ostjüdische Etappe 1916–18* [Hawdoloh and the Last Post: Memories of the Eastern Jewish Experience 1916–18]. Berlin: Jüdischer Verlag.

Hank, S. and Simon, H. 2002. *Feldpostbriefe jüdischer Soldaten 1914–1918: Briefe ehemaliger Zöglinge an Sigmund Feist, Direktor des Reichenheimschen Waisenhauses in Berlin* [Jewish Soldiers' Letters from the Field 1914–18: Letters from Former Pupils to Sigmund Feist, Director of the Reichenheim Orphanage in Berlin], Volume I and II. Teetz: Hentrich and Hentrich.

Heid, L. 1995. *Maloche – nicht Mildtätigkeit: Ostjüdische Arbeiter in Deutschland 1914–1923* [Hard Work – not Charity: Eastern Jewish Workers in Germany 1914–1933]. Hildesheim: Olms.

Kaplan, M.A. 1991. *The Making of the Jewish Middle Class: Women, Family and Identity in Imperial Germany.* New York: Oxford University Press.

Klemperer, V. 1996. *Curriculum Vitae: Erinnerungen 1881–1918* [Curriculum Vitae: Memoirs 1881–1918]. Berlin: Aufbau.

Kraemer, D.C. 2007. *Jewish Eating and Identity Through the Ages.* London: Routledge.

McIntosh, A. and Zey, M. 1998. Women as Gatekeepers of Food Consumption: A Sociological Critique, in *Food and Gender: Identity and Power*, edited by C.M. Counihan and S.L. Kaplan. London: Routledge, 125–44.

Marx, J. 1939. *Kriegs-Tagebuch eines Juden* [War Diary of a Jew]. Zürich: Verlag 'Die Liga'.

Maurer, T. 1986. *Ostjuden in Deutschland, 1918–1933* [Eastern Jews in Germany, 1918–1933]. Hamburg: Christians.

Plessner, A. ed. 1918. *Wie es Josef Kraft in der 'feindlichen' Synagoge erging und andere jüdische Feldpostbriefe aus dem grossen Weltkrieg* [How Josef Kraft Fared in the 'Hostile' Synagogue and other Letters from Jewish Soldiers in the Field during the Great War]. Berlin: publisher unknown.

Reichsbund Jüdischer Frontsoldaten ed. 1935. *Gefallene deutsche Juden: Frontbriefe 1914–18* [Fallen German Jews: Letters from the Front 1914–18]. Berlin: Vortrupp Verlag.

Salomonski, M. 1918. *Jüdische Seelsorge an der Westfront* [Jewish Spiritual Welfare on the Western Front]. Berlin: Louis Lamm.

— 1917. *Ein Jahr an der Somme* [A Year at the Somme]. Frankfurt a/Oder: Königliche Hofbuchdruckerei Trowitzsch & Sohn.

Scholem, B., and Scholem, G. 1989 *Mutter und Sohn im Briefwechsel 1917–1946* [Mother and Son in Correspondence 1917–1946], edited by I. von Shedletzky with T. Sparr. Munich: C.H. Beck.

Scholem, G. 1997. *Von Berlin nach Jerusalem. Jugenderinnerungen* [From Berlin to Jerusalem: Memories of Youth], translated by M. Brocke and A. Schatz. Frankfurt a/Main: Suhrkamp Verlag.

Segall, J. 1922. *Die deutschen Juden als Soldaten im Kriege, 1914–1918: Eine Statistische Studie* [German Jews as Soldiers during the War, 1914–1918: A Statistical Study]. Berlin: Philo-Verlag.

Seligsohn, H. 1903. Geschichte der Familie Seligsohn zu Samotschin, Provinz Posen: Tragische und humoristische Geschehnisse aus Samotschin [History of the Seligsohn Family in Samotschin, Province of Posen: Tragic and Humorous Events from Samotschin], unpublished memoir. Schönenberg, Leo Baeck Institute (LBI), New York, ME 596.

Simon, E. 1998. *Sechzig Jahre gegen den Strom: Ernst A. Simon, Briefe von 1917–1984* [Against the Current for Sixty Years: The Letters of Ernst A. Simon, 1917–1984], edited by the LBI, Jerusalem. Tübingen: Mohr Siebeck.

Specht, H. 2006. *Die Feuchtwangers: Familie, Tradition und jüdisches Selbstverständnis im deutsch-jüdischen Bürgertum des 19. und 20. Jahrhunderts* [The Feuchtwangers: Family, Tradition and Jewish Identity in the 19th and 20th Century German-Jewish Bourgeoisie]. Göttingen: Wallstein.

Straus, R. 1962. *Wir lebten in Deutschland: Erinnerungen einer deutschen Jüdin, 1880–1933* [We Lived in Germany: Memories of a German Jewess, 1880–1933]. Stuttgart: Deutsche Verlags-Anstalt.

Tannenbaum, E. 1915. *Kriegsbriefe deutscher und österreichischer Juden* [War Letters from German and Austrian Jews]. Berlin: Neuer Verlag.

Toller, E. 1978. *Gesammelte Werke 4: Eine Jugend in Deutschland* [Collected Works 4: A Youth in Germany], edited by J.M. Spalek and W. Frühwald. Frankfurt a/Main: Carl Hanser.

Wertheimer, J. 1987. *Unwelcome Strangers: East European Jews in Imperial Germany*. New York: Oxford University Press.

Weil, G. 1930. Vom jüdischen Volksheim in Berlin [On the Jewish People's Home in Berlin], in *Jüdische Wohlfahrtpflegeund Sozialpolitik* [Jewish Welfare and Social Politics], 281–9.

Ziemann, B. 2007. *War Experiences in Rural Germany, 1914–1923*, translated by A. Skinner. New York: Berg.

Zuckmayer, C. 2002. *Als wär's ein Stück von mir: Horen der Freundschaft* [As if it were a Part of Myself: Hours of Friendship]. Frankfurt a/Main: Fischer.

PART II
Home Front: The Citizens Adapt

Chapter 5

Food Provisioning on the German Home Front, 1914–1918

Hans-Jürgen Teuteberg

Introduction

In the last third of the nineteenth century Germany shifted from a traditional agrarian to a modern industrial state. A rapid rise of the urban population during this period increased not only the demand for foodstuffs, but at the same time led to a shift from vegetable to animal products. The food requirements of big towns could no longer be supplied from the surrounding area alone and additional foodstuffs had to be imported from more distant German regions or abroad. The German Empire's imports of agricultural produce nearly doubled between 1900 and 1912. The volume of grain imports, including animal feed, accounted for 25 per cent of all German imports in the last decade before the First World War.[1]

The dependence on foreign countries can be shown in the meat economy. With an expansion of life stock since 1871 peasants needed more and more animal fodder, since they could not produce a sufficient quantity by themselves. This made imports inevitable and each year until 1914 5m tons of concentrated feed were imported. Germany's dependence on import of fats, which stood at 40 per cent, was still higher. Germany produced most of its animal fat in the form of butter and output depended on the success of the life stock sector. Butter production required additional concentrated fodder. Therefore, even domestic fat production depended indirectly on foreign imports. In 1900 the German population consumed 30m tons of grain of which 7m tons were imported. It has been estimated that nearly one third of food and fodder consumed in Germany was imported before the World War.

This dependence on imports existed also in some other European countries. According to statistics published by the League of Nations Western Europe produced on average 37.8m tons of grain per annum between 1909 and 1913 and it imported 26 per cent from Eastern Europe and the USA. The German government pursued this international trade as far as possible, because large grain imports were imperative to protect Germany's large industrial exports.

Although the German government anticipated an economic blockade in the event of war, the state did virtually nothing to reduce the dependence on food

---

[1]   Skalweit and Krüger 1927: 11, Winckel 1915: 12–14, Steinkühler 1992.

imports. There was a general expectation that a future war would be similar to the Franco-Prussian War of 1870–71 which lasted only a short time. In this case German agricultural production could secure food provision. Around the turn of the century, some economists presented statistics to demonstrate the urgency of additional food imports and the need to establish national grain stores. A ministerial committee was established and the Economic Council of the *Reichstag* [parliament] discussed the question. Finally, the council decided to obtain expert opinion on the problem of import dependency. Regardless, all these plans did not come to fruition before the outbreak of war. These delays can be attributed to the following causes. First, it was not clear whether the central government or the German states should finance the grain stores. Secondly, there was considerable opposition to the prospect of destroying the liberal economy. Finally, these proposals were not implemented because the policy was perceived as undermining the government's next budget in 1915. All these factors prevented an economic mobilization in peacetime.

During the first months after the outbreak of war in August 1914 food provisioning did not present a major difficulty. There had been a very good harvest and initial hoarding of foodstuffs during the first weeks ended rather quickly. The official view still was that a British blockade could easily be resisted. Therefore, the only legislation passed in the *Reichstag* mandated new stock-taking of grain and flour.

The Foodstuffs Markets under Pressure from the Allied *'Hunger Blockade'*

Immediately after the beginning of the war on 5 August 1914 Britain declared a stop of all food transports to Germany and in November an economic blockade was imposed by all combatant nations allied with Britain. The German Empire designated this blockade illegal under international law, because ships of neutral states and civilians could be affected. After a short while Germany was severely damaged by the blockade because it had been the most successful exporter in Europe. The statistic showed that 75 per cent of German exports went to other European nations, which accounted for 57 per cent of its imports.[2] Therefore, the blockade was perceived as destroying the successful division of labour built up in the nineteenth century not only in Germany but internationally.

Through the loss of imports due the 'hunger blockade' – as the blockade was termed in Germany – a decrease of agricultural output began. With the loss of imports of animal feed the life stock sector in Germany declined by 65 per cent in the course of the war.[3] With imports accounting for one third of food supplies before the war, the food situation of 69 million Germans increasingly developed into a major crisis.

---

2   Eltzbacher 1914: 73ff.
3   Eltzbacher 1914.

Another important factor accounting for the decline in German agricultural production was the shortage of labour after the call-up of many peasants and their servants to the army. To compensate for the shortfall, the state deployed prisoners of war in agriculture as well as in other economic branches. At the end of the First World War about 900 000 prisoners of war were active as land workers.

Immediately after the outbreak of the war, a series of actions instituted a new food policy. At first, this was little more than an accumulation of acts and decrees, which were hastily issued and frequently remained rather ineffective due to inadequate cooperation. On 31 July 1914 all exports of life stock as well as animal products were prohibited and from 4 August import duties were cancelled. On 26 October legislation enabled towns with more than 10,000 inhabitants to fix maximum prices for all foodstuffs. This was intended to prevent profiteering. These new price regulations only affected retailers. A comprehensive rationing policy of all foodstuffs was at this time not planned, since it was generally believed that Germany would win the war.

The first maximum prices led to unequal supplies in the cities, because most produce went to those places where prices were higher or where there were no official price regulations as yet. Thus, maximum prices became minimum prices in reality. In contrast with grain, there was at first no fixed price for meat. Cereals were also used by the peasantry as animal feed and meat prices increased much faster as a result and the availability of meat and meat products declined. Like other foodstuffs meat partly disappeared from the normal distribution channels and it was sold directly to consumers which was illegal.

The Emergence of a National Nutrition Economy and Rationing Systems

In order to strengthen control from January 1915 onwards the stores of producers and consumers could be confiscated and a new *Reichsgetreidestelle* [National Grain Office] was established. This was the beginning of food rationing.[4] This new wartime body brought together producers and traders and the state participated by investing capital in the new institution. Following this pattern, most other foodstuffs became subject of official control and twenty similar corporations were established until 1917. A *Kriegsernährungsamt* [War Nutrition Department] which coordinated the different ration schemes was founded in 1916. Following this centralized model, a *Reichsverkaufsgesellschaft* [National Purchasing Department] was established. It had a trade monopoly and was responsible for importing grain, flour, rice and colonial products from neutral states and occupied regions. However, the German Empire had to compete on the world markets not only with their enemies, but also with its own allies. The volume of food imports from foreign countries, especially the Netherlands and Scandinavia, as well as from occupied regions remained small throughout the war.

---

4   Skalweit and Krüger 1927: 167–9.

The war economy was always dominated by two contradictory factors; viz. the army's inevitable requirements and constraints on the reduction of consumption on the home front. Planning of food provisioning therefore changed several times. The establishment of a war nutrition economy was initially based primarily on local government. This was due to the German tradition that local governments were responsible for feeding the poor as a part of their pre-industrial welfare work. In wartime the central government felt morally obliged to provide for all families of men who served in the army. Thus, traditional welfare policy was expanded to provision for many urban dwellers.

The first step in the establishment of a national nutrition economy was the bread and grain order of 25 January 1915. All German towns and rural districts took over the function of a wholesale dealer. The *Reichsgetreidestelle* at first provided for all local governments, except those which contained enough arable land to provide all food for their inhabitants. The towns purchased the necessary quantity of grain and transported it to nearby mills. From there the flour was distributed to urban bakers.[5]

In order to purchase and distribute such large quantities of foodstuffs the formation of new institutions was necessary. The authorities had to distribute special bread cards to each inhabitant who had a claim to this main foodstuff. At first uniform guidelines were absent and many new orders had to be issued. At a local level, the police, trade inspectors and slaughter houses as well as administrative departments for schools and welfare participated in the national food provision. Apart from local organization, special 'foodstuff offices' for bread, potatoes, meat and milk as well as for the distribution of ration cards, price inspection and control of illegal practices were established. Because local officials often did not have enough special knowledge about the foodstuffs they were in charge of, some undermined the implementation of food orders instead of cooperating. Generally, local communities' foodstuff purchases at the wholesale level increased considerably. From 1915 onwards, the urban offices had to distribute vegetables, fruit, milk and butter with purchase permits. Because demand for foodstuffs was always larger than supplies the rationing system was changed. Now each person should, in principle, get the same quantity without consideration of differences in age, gender or individual needs.[6]

Because of severe food shortages it was also necessary to build-up an adequate control system. This included propaganda in order to promote the rationing system for the common weal. Food rationing was no new phenomenon but on this grand scale it had no historical precursors. The new ration card gave no guarantee to quantities listed, with the exception of the bread ration. The control of retailers was very difficult, because it was impossible to scrutinize all transactions and sales.

---

5   Wiedfeldt 1919, vols. 50–3.

6   Skalweit and Krüger 1927: 196.

The nutrition economy, in principle, aimed at a fair distribution of food supplies and rations were graduated into several consumer groups. Apart from the men in the army, navy and heavy workers in war production, several other key occupations, children, pregnant women, the sick and old people received special rations. Only self-suppliers in the rural areas were not affected by food restraint. In the second half of the war more and more foodstuffs were rationed. From April 1916 onwards, not only bread and flour but also potatoes and, from May onwards, butter and sugar were rationed. Further in June meat was rationed and from November 1916 onwards also eggs, milk and other fats (i.e. margarine and edible oil). Apart from fruit and vegetables, the only food items not rationed were game, rabbits and horses.

Changes in the Consumption of Bread, Potatoes and Meat

Local governments rejected a proposal to introduce uniform rations throughout the German Empire, because the food habits in the individual states were rather diverse and, therefore, also the demand for many foodstuffs. Another factor why the rationing system was never absolutely precise was that households with moderate means exchanged their cards for more expensive foodstuffs like meat and eggs with coupons for cheaper bread or potatoes. As a result, these private transactions, which were not forbidden, flourished.

To ensure the safety of bread provision, which was essential, arrangements were made to stretch the existing stocks. This was necessary not only in view of the decline of agricultural workers, draught horses and imports, but also because of recurrent bad harvests. In such times, the extraction rate of rye for bread flour was increased. This bread was more nutritive but also not as tasty. Potatoes were also added to the flour, but these did not have the same degree of starch.[7] These policies introduced the first war-bread. Subsequently, there were other admixtures to the rye flour such as barley, oats, beans and peas as well as turnips. From 1917 other methods of stretching the rye flour were adulterations with straw or wood shavings. Experiments of producing bread with ground chestnuts, acorns, nuts or beech flowers were not very successful.

Rye flour had a high moisture content and it was liable to clump and acquire a musty smell if kept or transported for longer distances. Therefore, people rejected this war-bread, especially when it did not contain enough mashed potatoes.[8] After 1915, night baking was no longer allowed and the consumption of popular items such as rolls declined. Likewise, the sale of cakes was prohibited in 1917.

Considering the early grain shortages, mayors and municipal authorities urged the consumers to be sparing in bread consumption and other flour products. This also applied to butter and lard, but such appeals for voluntary restriction were

---

7　Roerkohl 1991: 95.
8　Skalweit and Krüger 1927: 34.

unsuccessful, because many retailers hoarded these commodities in the hope of greater profits later. They were subsequently condemned as a profiteering trade. By means of price controls, local governments managed to retard price rises with the exception of meat, butter and legumes until spring of 1915. Therefore, there was little talk about a reduction of living standards in Germany in the early part of the war. From March 1916 onwards price controls were extended to nearly all important foodstuffs. Food accounted for one third of a worker's household expenditure until the end of 1915 but at the end of the war this had increased to half because fixed prices had to be raised on several occasions. These waves of price increases affected households differently. Heavy workers in war industry received several wage increases, whereas craftsmen, white collar workers or low-ranking officials on salaried incomes suffered much more from rising food prices.

Contrary to expectations, potato supply also became 'a special problem child of war provision'.[9] Price regulations, of course, were influenced by changing supplies throughout the year. After the harvest in mid-October potatoes were traditionally stored in cellars until March of the following year. From April to June households depended on what peasants brought to the weekly markets. However, quantities were often limited in the expectation that the potatoes could be sold at a higher price as cattle feed. A national potato office was established by the government in 1915, but total rationing of potatoes was impossible. From April 1916 onwards free sales of potatoes in urban markets were forbidden but this prohibition was not successful. October 1916 marked the beginning of the first large potato famine. This was the result of a very bad harvest coupled with the collapse of imports from Austrian Galicia and Romania. For example, the weekly potato ration in Münster in January 1917 amounted to only three pounds per person.[10]

To compensate for the shortfall, people resorted to turnips. For the first time, this animal feed became a major foodstuff for human consumption. There were new recipes such as turnip soup, turnip meat balls, turnip dumplings, turnip schnitzel, turnip pancake and turnip pudding. Naturally, turnips were also added to vegetable dishes. The turnip became a daily staple and it symbolized hunger and misery.[11] In 1917 the turnip ration per person was cut to only one pound per week.[12]

The provision of meat, milk and fat clearly demonstrate that these foodstuffs had become central problems on the German home front. As supplies of potatoes fell short of expectations, the slaughter of animals and, especially, pigs increased. Following a decree of May 1915, which was later referred to as the 'pig killings', a third of the pig population amounting to between 1 and 2.5m animals were slaughtered immediately. This action was uneconomic and it led to a sudden rise

---

9   Schulte 1930: 322.
10   Schulte 1930: 262.
11   Oberschelp 1993: 317f.
12   Schulte 1930: 367.

of meat prices because many peasants sold the slaughtered pigs illegally directly to private consumers or to canning factories.

In view of this unsustainable situation, the cattle dealers association agreed with the newly established national meat office that it was necessary to introduce a heavily regulated meat economy.[13] In order to economise on consumption, the state introduced two meatless days a week for the urban population. Subsequently, the same restriction was introduced with regard to sale of fat. Local offices established their own facilities for fattening pigs which were sold directly to butchers.[14]

The supply of meat, of course, was linked to the production of animal fat such as lard and butter. Very different prices could result in confusion and arguments. A national fats office had been established in 1916, but it was impossible to close the gap between demand and supply with fixed prices alone. With the same aim restrictions on milk consumption were announced at the end of 1915. This meant that a household could drink no more than one litre of milk per day. From October 1916 onwards only pregnant women, babies and sick people could receive whole milk. Milk was available until 11 o'clock in the morning and subsequently the remainder was freely available. As a result of this policy, long queues of buyers became commonplace everywhere in the early morning in front of milk shops.[15]

By contrast, vegetables and fruit were never rationed during the war, because there simply too many varieties and supplies fluctuated with the harvest. Another problem was that these foodstuffs could spoil rather quickly. Apart from the use of turnips, there was an increase in consumption of processed foods such as artificial honey, soup cubes, pudding powder and coffee substitutes. All these new foodstuffs had been invented before the war, but they acquired much greater popularity.[16]

The production of food surrogates, primarily a so called 'war pap' made of turnips, was a profitable business because the ingredients were of inferior quality. There was also a sausage which contained five per cent fat and 80 per cent water. Another common item was cheap mushrooms.[17] After food chemists detected substances such as chalk, gypsum and ash in food substitutes in 1918, it was generally forbidden to sell them. Nevertheless, local offices knew that there were many food substitutes in Germany at the end of the First World War.[18] It was estimated that 12 per cent of households on small incomes regularly used these kinds of surrogates.

---

13   Lorz 1938: 392–5, Skalweit and Krüger 1927: 96–106, Aereboe 1927: 50, Burchardt 1974: 67.
14   Skalweit and Krüger 1927: 36–7, Roerkohl 1991: 16, Schulte 1930: 200, 206, 237.
15   Schulte 1930: 243.
16   Roerkohl 1991: 218, 221, Skalweit and Krüger 1927: 52, 58.
17   Burchardt 1974: 68.
18   Skalweit and Krüger 1927, Roerkohl 1991: 223, Schumacher 1915: 87–90.

The Great Provision Crisis of 1916 and its Economic and Social Aftermath

The scarcity of so many foodstuffs made shopping much more troublesome and laborious. From 1915 or 1916 onwards long queues in front of shops became part of daily life in German towns. First, the housewife had to obtain ration cards and purchase permits and then she had to get information about the time and place for sales and allowed quantities of foodstuffs from newspapers or placards. Some women and their children began queuing in front of a shop during the night. Often they returned disappointed because supplies in the shops were simply insufficient. The allocation of butter, margarine and eggs happened at different places and days of the week. Those who arrived too late no longer had a legal claim to their rations.[19] Lengthy waiting in queues frequently led to pushing, insults and even fights. Further, the preparation of daily meals was increasingly difficult at a time when food consumption was not oriented towards what people wanted but depended on the availability and quantities of rationed food. As a result, meals were simplified, many lacked taste and traditional food habits had to be dispensed with.[20]

The authorities were aware of this problem and the state developed nutrition propaganda to help housewives.[21] For example, special cookery books were published. Large numbers of women's associations and the 'National Women Service' promoted the spread of local war kitchens and they set up special cooking courses. These provided a good opportunity to talk about nutritive values and more rational kitchen work. Another purpose of the propaganda effort was to overcome prejudices against new foodstuffs and the dislike for horse meat. Propagated also offered the opportunity to promote healthy eating habits such as longer chewing. It was claimed that this helped to satisfy the appetite under the motto 'food chewing is half digestion'.

The reduction of food quality and quantity by means of rationing required increased expenditure of time and energy by housewives, but it also had the secondary effect of raising the status of women's domestic work. Wartime propaganda maintained that 'the rifleman's trench leads through the kitchen at home'. Furthermore, urban housewives were given detailed information on how to use leftovers and how kitchen waste could be used to feed life stock. And last but not least, another message in cooking courses was the instruction that 'the child has in any case to empty its plate'.[22]

The deteriorating domestic food supply situation resulted in the expansion of mass feeding through local war kitchens in the last two years of the war. These had several advantages. In the first place, they were a legal protection against exorbitant prices. There were no long queues and meals were cooked in a rational manner. There were several predecessors to war kitchens. Apart from school meals, there

---

19   Johann 1968: 325, Ullrich 1982: 95, Schulte 1930: 22, 240.
20   Daniel 1982: 218, Schulte 1930: 188, 212, 217, Follenius and Fassmann 1917.
21   Winckel 1915: 68–76, Roerkohl 1991: 179, 205–9.
22   Eifert 1985: 281–305, Winckel 1915: 68–76, Altheim-Gotheiner 1916.

were public soup kitchens, which provided relief for the poor, before the First World War. From the potato crisis of 1916 onwards, local authorities started to use these older organizations as war kitchens. Meals were prepared in a large central kitchen and served in smaller distribution places in the different urban districts. Households on small incomes, single persons or families without a breadwinner received one warm meal each day as a minimum. After 1916 the kitchens offered meals at a very low price to those who had a food coupon. Initially, the take-up of this new form of the mass feeding was positive, but subsequently it caused resentment against the middle classes because the meals were reminiscent of the tradition of feeding the poor. The Social Democrats called for a general extension of mass feeding to the entire urban population, but this suggestion of forcible feeding for all was rejected by the majority of the population.[23]

War kitchens distinguished between working class users, the middle classes and canteens for industrial workers. In the event, sometimes only fifty per cent of the inhabitants went to war kitchens, probably because this form of eating threatened to undermine people's social prestige. In the winter the two war kitchens in Münster were closed in contrast to those in industrial big cities.[24] The meals consisted mainly of a substantial soup, mostly consisting of potatoes and legumes, while meat and fats were only used sparingly. From the winter of 1915–16 onwards, more and more meals were made from turnips and food substitutes. Eating had become a cause of complaint because quality was very poor and the diet was monotonous. At this time children were seen begging for bread crumbs in bakeries and scouring the weekly outdoor markets left-over vegetables and fruits.

In the face of intensifying food shortages, private self-provisioning by the urban population increased during the last two years of war. The authorities appealed to the public to grow food in their gardens and also convert fallow land to food production. This war garden cultivation helped to make the daily menu more varied and to fill dietary gaps. Families from all social classes intensified their garden work and rents to lease land increased by about one hundred per cent.[25] Many families went to nearby forests to gather wild fruits and mushrooms. Others kept poultry, goats and rabbits to get more eggs, milk and meat. Home slaughtering of cows and pigs was forbidden in urban areas from 1900 and people sometimes boarded a larger animal with a peasant. From 1917 onwards, it was not allowed to keep these 'pension pigs' longer than 12 weeks. This form of self-provisioning was rather expensive and, therefore, only available to the wealthy. Despite increasing state regulation, there was still a substantial free food trade. Urban residents could ask a peasant or trader so sell them illegal foodstuffs which had been hoarded, often in large quantities. Many illegally purchased items were

---

23   Roerkohl 1991: 177, 230, 235, Skalweit and Krüger 1927: 40f, 44. On feeding the poor in Germany see Teuteberg 2009.

24   Schulte 1930: 145, 197, 214.

25   Winckel 1915: 48–60, Schulte 1930: 324.

luxuries or foodstuff which could be stored easily such as grain, rice, lentils, peas and preserves.[26]

Large waves of town people travelled to the countryside on trains on weekends to buy these foodstuff and also flour, bacon and fresh butter without food coupons. This practice could result in violence or plundering and crowds of urban people sometimes stormed a peasant's house, trampled down his fields and pulled fruits from the trees.[27] At the railway stations, the police searched people's luggage and confiscated illegal foodstuffs. Nevertheless, the state did not actually prohibit illicit hoarding because it had become necessary for survival in many urban households.[28] In the last year of the war it was estimated that about half of the registered quantities of meat, eggs and fruit, one third of milk, butter and cheese as well as a smaller proportion of flour, meat and potatoes were withdrawn from the legal food economy.

The Effect of the Wartime Diet, 1916–1918

The famine during the last two years of the war was not a general phenomenon affecting the entire population. On the contrary, it exacerbated deep inequalities and a new 'food hierarchy' emerged in German society during in this period.[29] From top to bottom, this hierarchy distinguished between the following groups:

- Food producers, traders and retailers,
- Private self-suppliers with compensation deals and the wealthy with links to the black market,
- Workers in war industries with relatively high wages and extra food allowances,
- Residents in small provincial towns with a substantial garden and direct contact to peasants in the surrounding countryside,
- Urban residents such as salaried employees, low-ranking officials, craftsmen and non-unionized manual workers,
- Families of soldiers living without a male bread-winner, and
- Single older people and inmates of institutions.

The First World War had a profound impact on the health of the population. Under nourishment began during the winter of 1916–17 and the real hunger catastrophe occurred in 1918–19. During the turnip winter many mayors certified that about 90 per cent of the inhabitants of their towns were concerned with the 'hunger problem'. The inmates of prisons were among the first group where continued

---

26   Roerkohl 1991: 262, 289, 295.
27   Daniel 1982: 225, 228f, Skalweit and Krüger 1927: 219.
28   Schulte 1930: 255, 280, 299f, 344.
29   Huegel 2003: 473–4, Rintelen 1932: 85–90.

under nourishment caused sickness, frequently resulting in premature death.[30] From 1916–17 doctors noted that 25 per cent of their patients had experienced a large loss of weight. This resulted in fatigue as well as lowered concentration and efficiency at work. Children at school showed unusual loss of attention.[31] Between 1914 and 1918 an excess of 160,000 people died of tuberculosis in the German Empire compared with the prewar period. This can be largely attributed to rising hygiene problems. The large wave of Spanish influenza throughout Europe in at the end of the war took at toll on public health and mortality rates, but was not directly caused by the famine.[32] Changes in food consumption between 1916 and 1918 are detailed in Table 5.1.

Table 5.1   Wartime Rations of Important Foodstuffs, 1916–18 (in per cent of the peace consumption)

Foodstuffs	1 July 1916 to 30 June 1917	1 July 1917	30 June 1918
Meat	31.2	19.8	11.8
Lard	213.9	10.5	6.7
Butter	22.0	21.3	28.1
Vegetable fats and oils	39.0	40.5	16.6
Legumes	24.2	–	6.6
Flour	52.5	47.0	48.1
Potatoes	70.8	94.2	94.3

Source: Burchardt 1974: 69.

In the last year of the war discontent with the food situation became a constant subject of conversation. A local newspaper made the remark: 'So in wind and weather are standing often hundreds of freezing, waspish and pale people waiting in a long queue pushing and grumbling about war, hunger, prices and hoarding'. The queues graphically demonstrated the profound gap between urban consumers and peasants. The flourishing black market, menus at high-class restaurants and

---

30   Roerkohl 1991, Schulte 1930: 274.
31   Thiele and Lorentz 1919: 7–37.
32   Huegel 2003: 473–587.

stocks in delicatessens shops demonstrated that there was much more food than the official supply stations had at their disposal. For the majority of the population, who were increasingly embittered, this was a failure of the state. Contrary to the first years of the war food rationing now resulted in social injustice. The hoarding, foraging and luxury feeding of a small minority were seen as a provocation by the general public because this was generally only within reach of the wealthy or upper classes.[33]

In Hamburg housewives, juveniles and children demonstrated for the first time with the demand for more bread. Some shop windows were damaged and bakeries plundered because the bread ration had again been shortened. Such disturbances were seen also in some other industrial cities. These protests and strikes by workers soon developed into political demonstrations which led to the revolution of November 1918 and the collapse of the German Empire.[34]

Conclusion

The end of the World War did not result in a sudden end of food shortages. Peasants' food delivery strikes, foraging, hoarding, black markets and the whole food rationing system persisted. On 12 June 1919 the British blockade was terminated and food imports slowly started again. Economic recovery was difficult throughout Europe and Germany no longer had sufficient foreign currency for large imports. In 1920 German agricultural output stood at only half its prewar capacity and domestic food production did not reach the 1913 level until 1928. Wartime food offices were gradually wound down and their staff was reduced.

On 19 January 1919 a new national assembly was elected and Germany's first president, the Social Democrat Friedrich Ebert, and his chancellors proclaimed not only a general peace and order but also a return to an efficient nutrition. The political upheaval should in no case disturb this reestablishment. But the food situation could not really be improved until 1923. This was due, in the first place, to the hyper-inflation which reduced wages, salaries and other income. In addition a large wave of unemployment reduced consumer demand. Only with the introduction of a new currency at the end of 1923 and the subsequent return to the free market system was it possible for Germany to gradually restore normal feeding. The intense political conflict in the new German republic and the programmes of the political parties demonstrated that securing the 'daily bread' was still seen as a main task.

---

33   Roerkohl 1991: 320.
34   Ullrich 1982, Burchardt 1974: 73, Scholz 1984.

References

Aereboe, F. 1927. Der Einfluß des Krieges auf die landwirtschaftliche Produktion in Deutschland [The influence of the war on agrarian production in Germany], in Wirtschafts- und Sozialgeschichte des Weltkrieges. Deutsche Serie [Economic and Social History of the World War: German Series], edited by J.T. Shotwell. New Haven: Yale University Press.

Burchardt, L. 1974. Die Auswirkungen der Kriegswirtschaft auf die deutsche Zivilbevölkerung im Ersten und Zweiten Weltkrieg [The impact of the war economy on the German civilian population in the First and Second World Wars]. Militärgeschichtliche Forschungen, 15, 65–97.

Altheim-Gotheiner, Elisabeth (ed.) 1916. Heimatdienst im ersten Kriegsjahr [Home Service in the First Year of the War], Jahrbuch des Bundes Deutscher Frauenvereine für 1916, Leipzig and Berlin: Teubner.

Daniel, U. 1982. Arbeiterfrauen in der Kriegsgesellschaft: Beruf, Familie und Politik im Ersten Weltkrieg [Working-class Women in Wartime Society: Occupation, Family, Politics in the First World War]. Göttingen: Vandenhoeck Ruprecht.

Eifert, C. 1985. Frauenarbeit im Krieg: Die Berliner Heimatfront 1914–1918 [Women's work in the war: the Berlin home front]. Internationale Wissenschaftliche Korrespondenz zur Geschichte der Arbeiterbewegung, 21, 281–305.

Eltzbacher, P. ed. 1914. Die deutsche Volksernährung und der englische Aushungerungsplan, Denkschrift [German Nutrition and the English Starvation Campaign, Memorial]. Braunschweig: Vieweg.

Follenius, R. and Fassmann, K. 1917. Der Zucker im Kriege [Sugar in the War]. In: Beiträge zur Kriegswirtschaft [Contributions to the War Economy], nos. 12–13, Berlin: Verlag der Beiträge zur Kriegswirtschaft.

Huegel, A. 2003. Kriegsernährungswirtschaft Deutschlands während des Ersten und Zweiten Weltkrieges [German War Nutrition Economy during the First and the Second World Wars]. Konstanz: Hartung-Gorre.

Johann, W. ed. 1968. Innenansichten eines Krieges: Bilder – Briefe – Dokumente [Interior Views of a War: Pictures – Letters – Documents]. Frankfurt a. M.

Lorz, F. 1938. Was wir vom Ernährungswesen des Weltkrieges nicht wissen! [What we do not know about nutrition during the World War], in Was wir vom Weltkrieg nicht wissen, edited by W. Jost and F. Felger. 2nd Edition. Leipzig: Fikentscher, 386–98.

Oberschelp, R. 1993. Stahl und Steckrüben [Steel and Turnips]. In: Beiträge und Quellen zur Geschichte Niedersachsens im Ersten Weltkrieg (1914–1918) vol. 1, Hameln: Niemeyer.

Rintelen, P. 1932. Deutschlands Bevölkerungsentwicklung, Nahrungserzeugung und Nahrungsverbrauch: Ein Beitrag zur Frage der Selbstversorgung des deutschen Volkes mit Nahrungsmitteln [Population Development, Food

Production and Food Consumption in Germany: A Contribution to the Question of the German People's Independent Food Supplies]. Münster: Fahle.

Roerkohl, A. 1991. Hungerblockade und Heimatfront: Die kommunale Lebensmittelversorgung in Westfalen während des Ersten Weltkrieges [Hunger Blockade and Home Front: Public Food Supplies in Westphalia during the First World War]. Stuttgart: Steiner Verlag.

Scholz, R. 1984. Ein unruhiges Jahrzehnt: Lebensmittelunruhen, Massenstreiks und Arbeitslosenkrawalle in Berlin 1914–1918 [An unsettled decade: Food riots, mass strikes and unrest among the unemployed in Berlin] in Pöbelexzesse und Volkstumulte in Berlin. Zur Sozialgeschichte der Straße, [Mobs and Riots in Berlin: A social history of the Street], edited by M. Gailus. Berlin: Verlag Europäische Perspektiven, 79–124.

Schulte, E. 1930. Kriegschronik der Stadt Münster 1914 [War Chronicle of Münster 1914]. Münster: Aschendorff.

Schumacher, H. 1915. Deutsche Volksernährung und Volksernährungspolitik im Kriege [German Nutrition and Nutrition Policy in Wartime]. Berlin: Heymann.

Skalweit, A. and Krüger, H. 1927. Die deutsche Kriegsernährungswirtschaft, in Wirtschafts- und Sozialgeschichte des Weltkrieges. Deutsche Serie [Economic and Social History of the World War: German Series], edited by J.T. Shotwell. New Haven: Yale University Press.

Steinkühler, M. 1992. Agrar- oder Industriestaat: Die Auseinandersetzungen über die Getreidehandels- und Zollpolitik des Deutschen Reiches 1873–1914 [Agrarian or Industrial State: The Grain Trade and Customs Policy in the German Empire]. Frankfurt a. M.: Lang.

Teuteberg, H.J. 2009. Historische Vorläufer der Lebensmitteltafeln [Historical precursors of nutrition tables], in Lebensmittel-Tafeln in Deutschland. Aspekte einer sozialen Bewegung zwischen Nahrungsmittelumverteilung und Armutsintervention, edited by S. Selke. Wiesbaden: Verlag für Sozialwissenschaften, 41–61.

Thiele, A. and Lorentz, F. 1919. Die Wirkungen der Hungerblockade auf die Gesundheit der deutschen Schuljugend [The impact of the hunger blocade on the health of the German school population], in Hunger! Wirkungen moderner Kriegsmethoden [Hunger: The Impact of Modern War Strategies], edited by M. Rubmann. Berlin: Georg Reimer, 17–36.

Ullrich, V. 1982. Kriegsalltag: Hamburg im Ersten Weltkrieg [Daily Life in Wartime: Hamburg in the First World War]. Köln: Prometh Verlag.

Wiedfeldt, O. 1919. Die Bewirtschaftung von Korn, Mehl und Brot im Deutschen Reiche – ihre Entstehung und ihre Grundzüge [The Rationing of Grain, Flour and Bread in the German Empire: Its Beginning and Basic Concepts]. In: Beiträge zur Kriegswirtschaft, nos. 50–53, Berlin: Hobbing.

Winckel, M. 1915. Kriegsbuch der Volksernährung [War Diary on Popular Nutrition]. München: Gerber.

Chapter 6

Bread from Wood: Natural Food Substitutes in the Czech Lands during the First World War

Martin Franc

Introduction

The Austro-Hungarian Empire entered the First World War inadequately prepared; it had insufficient food stocks for ensuring the long-term nourishment of both civilians and the army. This was partly a result of the conviction that the war would end shortly, but it was also combined with a belief that Hungary would be a reliable source of agricultural products.[1] Before long the monarchy realized that it would have to give more attention to the problem of food and would also need to accelerate its efforts in finding a solution. Successive measures were applied to the management and distribution of many commodities and also certain interventions were made in the production of some items.

One strategy used in managing the lack of produce was substitution, an experiment that began shortly after the outbreak of the conflict. There were two kinds of substitutes: synthetic and natural.[2] On the border between synthetic and natural substitutes, are those substitutes which were derived from natural sources, but were produced during this period using different and previously unknown technological procedures. The use of natural substitutes for many commodities was both more rapid and widespread. In many cases, the final product was not a complete substitute but a dilution of the original product through the addition of a substitute. In the case of natural substitutes, these were products made in the Austro-Hungarian Empire, as imports were too difficult to obtain. The most commonly applied substitutes were those made from various cultivated plants; experiments using wild plants were unsuccessful given consumer aversion and problems with collection.

---

1 At the beginning of the war, Czech agro-chemist Julius Stoklasa was very optimistic about the nutritional self-sufficiency of the Austro-Hungarian Empire. His confidence led to a conflict with Czech agrarians, particularly Karel Viškovský. See Viškovský 1915, Stoklasa 1916: 21. An assessment of the entire argument can be found in Sedivý 2001: 248–9.

2 The term 'substitute' was deemed commercially unsuitable and thus the word 'war' was used instead, e.g. war coffee. See *Národní politika* 3/4/1916: 6, Řeháček 2007: 289.

We can consider as a substitute those products that were recommended as alternatives to scarcer commodities and were treated in a way which made their taste resemble the original product. For example, smoked sea-fish was promoted by some companies as a substitute for smoked pork, which was in short supply. At the same time, there were efforts to use some of the less frequent or culturally taboo kinds of meat, for example, rabbit and horse meat. In this chapter, I will be focusing on the problem of natural substitutes in the civilian sector of the Czech lands, their perception by Czech speaking inhabitants and experts. The German speaking population of the Czech territories may have had quite different opinions and responses, but these are outside the remit of this analysis.

Substitutes for Bread and the Debate between Czech Scientists

Probably the greatest use of natural supplements was in bread flour. Mixing substitutes into commercially made bread had started before the war, and by 1916 the proportion of substitutes in flour was, according to official statistics, as high as 60 per cent.[3]

According to Julius Stoklasa, the contemporary expert, substitutes for bread flour were split into five groups. The first group included the flours from other field cereals and cultivated products (barley, oats, corn, Indian millet, rice, chestnuts), followed by malt products (malt flour, brewery grains), potato products, and starches and sugars (potato starch, tapioca, sugar).[4] In the last group he included other, until then, unusual ingredients, for example straw, hay, saltbush, clover, nettle (*Utrica divica*), finely ground wood or tree bark, island lichen (*Cetraria islandica*), water chestnuts (*Trapa natans*) and horse chestnuts (*Aesculus hippocastanum*).[5]

It was quite common to dilute flour with sugar. The Czech lands led in the growth and production of sugar beet in this period, but by the end of the war this too had become scarce. Flour production was increased through the common practice of very high extraction, at a rate of more than 80 per cent. The most widespread substitute was barley flour, while corn flour or potato additives were common. The majority of the substitutes from the fifth group remained at the research stage, partly due to the fact that the problem of collecting wild plants was never successfully resolved.[6] However, even the initial experiments caused angry reactions, a response determined by the Czech population's attitude towards the war. Even though the majority of drafted men fought loyally for the Austro-Hungarian Empire, the Czech population was not overly sympathetic

---

3   See Franzl 1915–16: 618, 750–51, Franzl 1916–17:102–3 for details of changes in state directives regarding substitutes.

4   This group was created for its specific connection to the brewing industry.

5   *Lidové noviny* 6/3/1917: 5, Stoklasa 1916: 23

6   Groups of school children and soldiers helped to collect these plants, *Lidové noviny* 3/18/1917: 4.

to the Habsburg monarchy and its interests were alien to them. Moreover, the German speaking population regarded the war as a means to achieve a closer bond between the Austro-Hungarian Empire and its war ally Germany; this was in sharp contrast to the interests of most Czechs. During the war, many measures taken by the Austrian state were accepted somewhat indignantly by the Czech population, especially when the measures were inspired by Germany.

In general, substitutes were considered a German invention, or more accurately, they were considered an invention of the lunatic minds of German scholars and they were ridiculed accordingly by the contemporary Czech press. People who joined the war research effort in support of the Austro-Hungarian Empire were subsequently regarded as traitors to the nation or at least as traitors to its people. Even the most important Czech researcher in the area of bread supplements, Julius Stoklasa (1857–1936), was later derided for this work.[7] During the war he was an Imperial Councillor, a professor at the Czech Technical University and vice-president of the Imperial and Royal Institute of Experiments of the Ministry of Public Works. After ascertaining that the agricultural production of the Austro-Hungarian Empire was insufficient for fighting the war, he started to experiment with and evaluate various potential substitutes. He used not only physiological and biochemical arguments, but in line with contemporary practice he often cited historical arguments in support of his recommendations. Stoklasa, who promoted the use of substitutes throughout the war, was later despised for his research activities mostly by those on the left of the political spectrum, who accused him of trying to feed people with straw.[8] In the memorial volume printed on his death in 1936, some of his colleagues and friends felt the need to defend his research with substitutes and expressed doubt that he ever promoted the use of straw or lichen as a flour substitute.[9]

The leading Czech authority in the field of nutrition, Eduard Babák (1873–1926), acted with greater care and stressed his opposition to the use of more extreme substitutes. In his publication *Nutrition of Plants* 1917, Babák doubted the practical value of Max Rubner's research into the possibility of nourishment using well-ground powder from birch wood. He argued that a higher level of potato or corn flour in bread flour lowered the nutritional value of the bread.[10] Similar opinions relating to the addition of straw and beef blood to bread were expressed in a note to the *Česká Revue*, probably by the botanist

7   Influential agrarians declined to intervene in Stoklasa's favours, Beran 1937: 397. A blow to Stoklasa's prestige came when, in 1915, he was not named the head of the Cereal Institute, which directed the distribution of cereal. Karel Visnovsky, his great rival was given the office instead, Chocenský 1937: 565–6.

8   After the formation of Czechoslovakia he was forced to leave public life. Later he became the vice-president of the Czechoslovak Agrarian Academy, but the path to the president's office was barred to him, Chocenský 1937: 567–8.

9   Beran 1937: 397.

10   Babák 1917: 39, 46.

Bohumil Němec.[11] Physiological and chemical research was not the only approach adopted; it was common that in the area of natural substitutes, popular traditions or the references to the past were also employed. In many cases it was these historical reminiscences or traditions that prevented the branding of substitutes as 'foreign innovations'. A rival of Stoklasa, Karel Viškovský did not accept that barley wheat was an innovative supplement, claiming that it had frequently been added to bread in Czech villages earlier in the nineteenth century. Therefore, its use was not innovative, merely a return to a practice of their ancestors.[12]

As early as 1915, botanist Karel Domin published a study with the telling title *Crab Grass: The Forgotten Czech Cereal*. In it he summarized all the historical and geographical uses of *Panicum sanguinale* and the closely related *Glyceria fluitans* and supported its partial reintroduction. Čeněk Zíbrt (1864–1932), a cultural historian, supported the social and historical arguments to an even greater degree.[13] He returned to a time in history when substitutes were most widely used: 1817, the era of the last major famine in the Czech lands, reflected in the title of his book, *Czech Cuisine in the Time of Shortage One Hundred Years Ago*. Zibrt did not propose a return to the humble lifestyle of previous generations, but wanted to illustrate the possible lessons that might be learnt from history and apply them to the acute problems of his day. The wide variety of substitutes came mostly from official recommendations and from the influence of the Patriotic Economical Society, founded in the custom and practices of the time. Flour made from the roots of dog grass and navew was mentioned, in addition to that made of clover, lichen and horse chestnuts. Zíbrt and other authors were fully aware that during the war, the Czech media were harshly critical of these 'uncultivated' substitutes. For example, when an article in an Austrian newspaper reported that the procedure for the production of bread from wood had been patented, *Lidové noviny* reacted with exceptional skepticism: 'Is it possible that according to the patented procedure tasty (?) solid (??) and consumable (??) bread, made of about 70–80 per cent of the current war-time flour and 20–30 per cent of wood will be produced?' The question marks were added by the editor's office and the piece ended with the scornful comment 'After the production of bread from clover – a bread from wood ...'.[14]

Between the lines it was evident that the preferred option to bread made from unacceptable substitutes was a universal reduction in the state-guaranteed rations, but the Austro-Hungarian monarchy could not bring itself do this.[15] However, especially at the end of the war, more and more frequently bread of any kind was lacking. The supplements brought only partial relief as the wildest ideas remained in the experimental phase. Some of the supplements had a negative impact on

---

11　Bohumil Němec 1915a: 631.
12　Viškovský 1915a: 9.
13　Domin 1915.
14　*Lidové noviny* 5/4/1917: 5.
15　*Lidové noviny* 5/30/1917: 5.

the health of consumers, causing digestive problems. They also damaged the mill equipment that was unused to these unusual raw resources and in addition there were problems with storage.[16]

Natural Substitutes for Coffee

The supplementation of bread flour was probably the most visible feature of the substitute industry in the Czech lands during the First World War, but this was not the only aspect. One specific development was in the field of another widely consumed product: coffee. Here the war resulted in the use of new substitutes for the old substitutes. Before 1914, poorer inhabitants used various coffee substitutes, mostly chicory, but also rye, figs and dried fruits.[17] Now, even these substitutes were unavailable and others had to be found. For example, in the case of chicory the amount of chicory root was lowered; its supply was centrally regulated by the state from 1915 onwards and dried sugar beet was added to the mixture. Other supplements used were acorns, fruit grounds, potato peelings and roasted beans. Real coffee beans were diluted with sugar.[18] Tea practically disappeared from tables, but it was not as widely used at this point. It was substituted mainly by dried leaves, mostly those from blackberry and raspberry bushes.[19]

Natural Substitutes for Beer

A much more significant role for substitutes was in beer production, which, after bread, was the second most important area of substitutions. The war caused an extreme intervention into the production of breweries, which had been in sharp decline.[20] Most of the trouble was caused by the severe shortage of barley, which was primarily being used as a substitute in the production of bread. The production of barley malt now involved the substitution of other plants containing starch or sugar as well as regular barley. Corn, oats, rice, Indian millet, *Pisum*, potato flour, potato starch, dried sugar beet, dog-grass, salt bush etc. were all used. For reasons of taste, the use of such substitutes had to be balanced by the addition of salt.[21]

In special cases, precious foodstuffs were employed as substitutions; for example, one brewery in northern Bohemia used honey imported from Hungary in place of malt. Some of the substitutes developed in the First World War continued to be used after the war because they were found to have beneficial properties.

---

16   Vilikovský 1936: 527–8.
17   *Lidové noviny* 5/29/1917: 4.
18   *Lidové noviny* 3/15/1917: 3.
19   Řeháček 2007: 299.
20   Vilikovský 1936: 696
21   Ibid.: 696.

For example, diluting malt with rice proved to be good for the production of the very light Silesian beers and adding dried sugar beets was helpful in keeping the head on the beer. The use of wild plants as substitutes caused similar negative reactions as to those to their use in bread, especially when they influenced the taste of beer. This was the case of beer made with *Agropyrum repens*, which according to the historian Vilikovsky 'had a characteristic taste of iodoform and that was the reason it disappeared after the initial experiments'. In the brewing industry, historical practices were also considered, for example, beer made from oats which had been done at the time of the Napoleonic Wars.[22] The most common substitute was regular beer watered down with the brewed hops which had been saturated with carbon dioxide; this resulted in a brew of approximately one degree beer.[23]

Natural Substitutes for Meat

It was quite common to find substitutes for meat as well. Experts recommended that meat consumption should be reduced, while at the same time they questioned earlier calculations regarding the required amount of protein per person per day. These conclusions were, in the main, shared by Julius Stoklasa, Eduard Babák and another important expert in the field of physiology, František Mareš (1857–1942).[24] Mareš added to the criticisms of the squandering of meat and to recommendations for reductions in consumption. Such views echoed folk heroes, for example a character in Jaroslav Hašek's *Good Soldier Švejk*, an old shepherd who pronounced, 'In recent years they devoured pork, poultry, everything soaked with butter and fat. Then God got angry at their pride and they will be humbled again, cooking weeds as they did in the time of the Napoleonic Wars.'[25]

The Habsburg monarchy introduced meat-free days, but in an interesting exception, permitted the consumption of certain unpopular meats on such days.[26] The substitution of some common kinds of meat with those less frequently eaten or previously taboo meats became normal practice. The war cook-book recommended, for example, the use of beef entrails, which were a waste product from the canning of beef flesh for the soldiers.[27] In the latter part of the war, there was a move to preserve milk-producing cattle and introduce other substitutes for meat, such as vegetables, cheese, eggs, smaller cattle, fish, poultry and game.[28] However, the prices of these commodities rose sharply because they were as

---

22   Ibid.: 697–8.

23   Ibid.: 697. Other substitutes were used too, such as ginger beer, Řeháček 2007: 299.

24   Stoklasa 1916: 11, Babák 1917, Mareš 1915.

25   Hašek 1954: 261, Mareš 1915: 86.

26   Franzl 1916–17: 50.

27   Urban 1915: 5.

28   *Lidové noviny* 3/9/1917.

unavailable to the general population as the standard meats. It was also possible to purchase horse or donkey meat in the shops.[29] At the fringes of society there was an increased consumption of what had been until then highly taboo meats, such as dogs, water-rats and the meat from carcasses.[30] At the other end of the spectrum, there were changes in the meat consumption of the aristocracy. For example, pork appeared for the first time on the menu of one of the most prominent Czech noble families, the Schwarzenbergs.[31] In the case of meat substitutes there were similar efforts to seek out the substitutes of the past. Zíbrt in his remark about 'times of shortage' mentions the consumption of frogs, snails, squirrels, and home grown turtles. In this instance, he was not referring to the famine of 1817, but the fasting seasons of even older times. His experience as an ethnographer was significant and he mentioned the 'strange animal delicacies' of rural children in Moravian Slovacko which included sparrows, mudfish and, pouch marmot's liver.

Ocean fish was recommended as a substitute for meat products, especially those with a high fat content; a previously rare item in the Czech diet.[32] Mushrooms were commonly presented as a possible substitute for meat, indeed they were recommended by Stoklasa. Babák was of a different opinion and rejected this substitution, citing their low nutritional value and difficulties in their digestion.[33]

In addition to the substitution of meat itself, it was necessary to devise substitutes for other meat by-products. The Austrian Minister of Nutrition, Hofer, promoted a product called 'war smoked meat' which was made from the blood and entrails of the 'inferior cattle from the occupied areas' together with legumes and, in their absence, barley and millet.[34] Blood was considered one of the most important substitutes for proteins but the Austrian administration had a problem with determining how exactly it should be used. Another option, which was used only to a limited extent, was the consumption of yeast-plants, again recommended by Stoklasa.[35] Němec also mentions a substitute for meat made from legumes and wheat flour, but he rejected it at the same time.[36]

Natural Substitutes for Fats

The supply of fats, for both human consumption and technical use was another major problem. The authorities introduced a fat-free day and other measures to

---

29   *Lidové noviny* 4/20/1917: 5.

30   *Lidové noviny* 3/19/1917: 2. Water-rat was very popular, ibid. 3/26/1917: 2, *Lidové noviny* 4/8/1917: 5, 9/15/1917: 2.

31   Besides pork, mutton also appeared more frequently, Cichrová 2003: 52, 61.

32   Stoklasa 1916: 12.

33   Babák 1917: 51, Němec 1917.

34   *Lidové noviny* 2/21/1917: 4.

35   Stoklasa 1916: 14–15.

36   Němec 1916a: 632. The outcome was described as 'virtually worthless.'

lower the consumption of fried food.[37] They also explored the production of oil from corn sprouts, grape stones and poppy seeds.[38] However, the main producer of poppy seeds was Romania with whom the Austro-Hungarian Empire was at war, thus the supply was curtailed.[39] In the case of fats, those wild plants rich in oil were considered as alternative supplies, for example wild radish or field mustard.[40] The substitutes for butter were unusual. Němec mentions, in addition to margarine, the dilution of butter with flour or substitutes made from skimmed milk and raw mashed potatoes.[41] Changes such as these are, in commercial terms, on the brink of falsifying the foodstuff itself. This is particularly the case when considering the addition of water to milk, an act which has been one of the most commonly practiced criminal strategies of adulteration for many centuries.[42] There are few instances of attempts to substitute milk and cheese with products made from soy, although a few households produced soy cheese.[43] However, as this plant was not grown in the Czech lands, its importation in war conditions was close to impossible and such experimentation generally remained in the laboratory testing phase.[44]

Natural Substitutes for Vegetables

The largest area of substitutions was that of vegetables.[45] In particular, swede turnip and navew were substituted for potatoes. The use of wild plants instead of cultivated vegetables was quite strongly opposed by the general public. One unnamed German professor from Bonn faced a strong critique from *Lidove noviny* when his recommendations in the *Leipziger Volkszeitung* were reported. He proposed the substitution of vegetables with the roots, leaves and stalks of thistle (*Carduus nutans*); roots and shoots of burdock (*Arctium lappa*), nettles and salt bush. The Czech newspaper was sharply ironic and in addition to other criticisms wrote, 'Whether this man has undertaken this diet is not known but there is no one who would envy him'.[46] On the other hand even the local commentators devised broader menus, using different kinds of wild plants and their experiments did not meet with the rejection experienced by their German counterparts. The

---

37   Řeháček 2007: 294.

38   Franzl 1915–16: 108.

39   Edvard Babák drew attention to the fact that there are a minimal number of plants high in oil content in the Czech lands. MÚA AV ČR, f. Edvard Babák, k. 3, inv. č. 219. He recommended the higher consumption of carbohydrates.

40   Řeháček 2007: 299.

41   Němec: 1916a: 632.

42   Regarding the falsifying of food and substitutes, see Dudeková 2007: 367.

43   *Lidové noviny* 4/23/1917: 3.

44   Vilikovský 1936: 855.

45   Řeháček 2007: 292.

46   *Lidové noviny* 3/20/1917: 4.

most far-reaching suggestions were those of Čeněk Zíbrt, but similar ideas were published by the chemist and keen populizer of science Bohumil Bauše (1845–1924).[47] The strategy of substituting common vegetables with wild plants did not appear feasible in commercial mass use, but the suggestions did perhaps inspire individual housewives.

Conclusion

This relatively short summary indicates the characteristic problems of finding natural supplements in the Czech lands at the time of the First World War. Above all, we can state that the relatively small community of Czech scientists was actively involved in the quest for suitable substitutes for those foodstuffs in short supply. At the same time, it must be stressed that historical arguments for such changes were also employed, in many cases by the scientists themselves. This was particularly the case in their efforts to overcome the distrust of conservative consumers, where it was pointed out that such substitutes were not foreign innovations but simply a return to the traditions of their ancestors. The most important source of inspiration was the experience of the Napoleonic Wars and the famine of 1817, but sometimes even older references were used, dating from the Middle Ages. The search for substitutes carried with it a nationalistic aspect. Germany was regarded as the home of substitutes, thus Czech supporters of substitution faced the possibility of being accused of a lack of patriotism, as happened to Julius Stoklasa. In addition to this nationalistic aspect we can observe a tendency to see the substitutes as the products of 'lunatic scholars'. The combination of these two factors resulted in the widespread stereotype of the dehumanized 'German professor'.[48]

The wide variety of substitutes suggests the breadth of research and interest from factions within the authorities who supported such activities. It is also evidence of other features relating to the supply of civilian food in the Czech lands. A certain common denominator of the distribution processes was its accompanying chaos and the frequent and sometimes thoughtless changes of traditional patterns of consumption. These changes were often ignored, especially by those from the upper classes of society. It was rare for any new war measure to be introduced in a timely fashion and gain the full support of the population, and this applied to substitutes as well. We can conclude that in reality, their contribution to food supplies was very limited and that their introduction, in many cases, merely further aggravated the population. Greatest importance was attached to the areas of bread flour and beer production. By the end of the war, the use of substitutes, together with the strict centralization and regulation of the distribution of food, could not prevent the revolts caused by the lack of food in many Czech towns. However, it

---

47   Bauše 1916: 1.

48   For example, news about efforts to replace beet sugar with maple syrup was described by *Lidové noviny* as 'another "invention" of the German professor': 6/12/1917.

appears, that despite the resentment and failure of many state efforts, there was no case of civilian starvation even during such a desperate time.

References

Anonymous. 1916–17. 'Z čeho možno ještě dnes vařiti pivo?, [Of what it is now possible to brew beer] *Česká revue*, No. 2, 119–20.

Babák, E. 1917. *Výživa rostlinami*. [Alimentation by the plants] Praha: F. Topič.

Bauše, B. 25/4/1916. Jarní žeň [Spring harvest], *Národní politika*.

Beran, R. 1937. Prof. Stoklasa a naše agrární politika, [Prof. Stoklasa and our agrarian policy] in *Památce profesora Julia Stoklasy*, edited by E. Reich and B. Vláčil. Praha: Československá akademie zemědělská, 394–9.

Chocenský, K. 1937. Julius Stoklasa jakým se mně jevil a jakým snad byl, [Julius Stoklasa how he appeared to me and what he perhaps was] in *Památce profesora Julia Stoklasy*, edited by E. Reich and B. Vláčil. Praha: Československá akademie zemědělská, 554–68.

Cichrová, K. 2003. Pozvání ke schwarzenberskému stolu na přelomu 19. a 20. Století, [Invitation to the Schwarzenberg's board at the turn of 19th and 20th century] in *Dobrou chuť* edited by L. Nikrmajer and J. Petráš. České Budějovice: Jihočeské muzeum, 51–61.

Domin, K. 1915. Rosička krvavá: zapomenutá česká obilnina, [*Panicum sanguinale*: the forgotten Czech cereal] *Časopis musea království Českého* 89, 47–94.

Dudeková, G. 2007. Stravovanie a zásobovanie města za prvej svetovej vojny, [Eating and food supply of the town at the time of the WWI] in *Dobrou chuť, velkoměsto. Documenta Pragensia XXV*, edited by O. Fejtová, V. Ledvinka and J. Pešek. Praha: Archiv hl. města Prahy – Scriptorium, 343–72.

Franzl, K. 1915–16. Mimořádná opatření hospodářská za války, [Emergency economic measures in wartime] *Česká revue*, 38–49, 103–11, 171–83, 236–47, 307–15, 370–78, 425–34, 491–500, 559–69, 618–27, 685–94, 748–58.

Franzl, K. 1916–17. Mimořádná hospodářská opatření za války, [Emergency economic measures in wartime] *Česká revue*, 44–50, 100–11, 225–30, 285–93, 338–45, 410–18, 479–84, 542–4, 603–8, 666–72.

Hašek, J. 1954. *Osudy dobrého vojáka Švejka za světové války*. [The Fateful Adventures of the Good Soldier Švejk during the World War]. Praha: Naše vojsko.

Mareš, F. 1915. *Výživa člověka ve světle fyziologie*. [People's alimentation in the light of physiology] Praha: J. Otto.

Masarykův ústav a Archiv AV ČR, v. v. i., f. Edvard Babák. [Masaryk's institute and Archive of the Czech Academy of Sciences].

Němec, B. 1915. Umělá výroba bílkovin, [Artificial production of the proteins] *Česká revue*, No. 1, 54.

Němec, B. 1915–16. Válečné náhražky, [War substitutes] *Česká revue*, No. 10, 631–2.

Němec, B. 1916–17. Zužitkování hub, [Utilization of mushrooms] *Česká revue*, No. 6, 352–3.

Řeháček, K. 2007. Dopady války na každodenní život obyvatelstva Plzně a okolních obcí (1914–1918), [Impacts of the war on the everyday life of inhabitants of Pilsen and surrounding villages (1914–1918)] in *Minulostí Západočeského kraje 42/1*. Ústí nad Labem: Albis international, 277–303.

Stoklasa, J. 1916. *Výživa obyvatelstva ve válce!* [People's Alimentation during the War] Praha: vlastním nákladem.

Šedivý, I. 2001. *Češi, české země a velká válka 1914–1918.* [Czechs, Czech Lands and the Great War 1914–1918] Praha: Nakladatelství Lidové noviny.

Urban, G. 1915. *Österreichisches Kriegs-Kochbuch.* [Austrian War Cookbook] Wien: Stefan.

Vilikovský, V. 1936. *Dějiny zemědělského průmyslu v Československu.* [History of the Agricultural Industry in Czechoslovakia]. Praha – Brno: Ministerstvo zemědělství republiky československé – Novina.

Viškovský, K. 1915. *Boj o chléb.* [Struggle for Bread]. Praha: A. Reinwart.

Zíbrt, Č. 1917. *Česká kuchyně za dob nedostatku před sto lety.* [Czech Cuisine in the Time of Shortage one Hundred Years Ago] Praha: A. Neubert.

Chapter 7

Hunger and Misery: The Influence of the First World War on the Diet of Slovenian Civilians

Maja Godina Golija

Introduction

For Slovenes, the First World War marked a turning point not only politically, but also culturally and socially. Slovenes actively participated or were indirectly involved in all stages of the war as soldiers, prisoners of war and volunteers. Civilians also played a role during the conflict as rebels, refugees, and internees. One of the most crucial, and most deadly, battlefields was situated along the Isonzo river in the western part of present-day Slovenia. The conflict permeated the lives of Slovenes on several levels. On the one hand it signified military operations, especially along the Isonzo Front, and on the other, it affected many aspects of civilian life.[1] This chapter will examine wartime life among the civilian population in Slovene towns and rural settlements and their daily battle to obtain enough food to survive. Battles were not confined to the trenches and were fought not only by the soldiers, smaller, but by no means lesser battles occurred away from the main front. The role and the suffering of the civilian population have largely been neglected in the texts written about this war. The exact number of civilian casualties has never been properly established. Estimates cite between two and thirteen million in Europe overall, most of whom were women and children. In Italy the number of civilian deaths roughly equalled that of the fallen Italian soldiers during the same period. No estimates have been made for Slovenia.

Slovene towns had a varied social and ethnic structure. During wartime their inhabitants suffered because of the lack of basic necessities. In addition, those Slovenes, suspected of being insufficiently loyal subjects of the Austro-Hungarian Empire, were subjected to strict control by the authorities and prohibited from displaying Slovene symbols, and carrying or flying the Slovene flag. Those who publicly voiced support for the Serbs, or were in favour of establishing a separate political entity for the South Slavs, were punished.

In addition to political pressure from the Austrian authorities, life in towns was disrupted by the influx of great numbers of mobilized men. Whether soldiers

---

1   Svoljšak 1993: 263.

on the move, the wounded, or convalescents, they all served to completely alter the towns' usual pace. Many public buildings were converted into makeshift barracks for troops or into hospitals, and military graveyards sprung up next to town cemeteries.

It could be argued that after the initial enthusiasm had abated, the war brought mostly confusion and economic uncertainty to the bewildered and largely unprepared Austria-Hungary. By March 1918, 60 per cent of males between 18 and 35 had been mobilized, causing a shortage of manpower in industry and agriculture. The authorities established a war economy. In an attempt to alleviate labour shortages, women were employed to do work that prior to the war had been done mostly by men. Their earnings, however, were much lower, and it was generally assumed that the war would be brief and soldiers would return and resume their jobs shortly.

The Shortage and Supply of Goods

By far the most pressing problem in Slovene towns was the shortage of basic necessities: food, clothes, fuel, and hygiene supplies. August Reisman, a lawyer from Maribor, wrote in his memoirs that queues were frequent, particularly at grocery stores, butcher shops, and bakeries – and, before that, outside the ration card offices. Despite the introduction of food ration cards, the supply of goods was still disrupted, which resulted in a shortage of food.[2]

In anticipation of a brief and victorious war, the Austrian authorities had largely neglected the question of food supply and distribution, but as the war continued from month to month, it became clear that this was not to be. The Austrian food supply depended on imports from abroad as well as from the Hungarian part of the monarchy. Instead of devising a carefully considered resource supply strategy, the government took only short-term measures on the food question which merely adapted to the existing situation without planning ahead. It established a number of centres that managed and supervised the supply of staples. These food distribution centres, each in charge of its own area, were organized on a local, usually municipal, level. Together with their neighbouring municipalities, the towns of Celje, Maribor, and Ptuj each represented a municipal food supply district managed by municipal economic offices.[3]

At the start of the war the authorities had moved to limit the purchase of certain goods, and the next step was the introduction of a food rationing policy. The government issued food ration cards or ration books with coupons, which allowed the holder to buy a restricted quantity of foodstuffs in a limited time. Ration cards for flour and bread, the first items to be rationed, were issued in April 1915. A holder was allowed to buy no more than 1,400 grams of flour and 1,960

---

2   Reisman 1939: 54.
3   Himmelreich 1998: 5.

Steiermark.

K. k. steierm. Statthalterei.

AUSWEIS

über den Verbrauch von Brot und Mehl in kleineren Ortschaften für die 102. und 103. Woche. 18./3. bis einschließlich den 31./3. 1917.

Menge für 14 Tage:

Allgemeiner Ausweis 3 kg 92 dkg Brot oder 2 kg 80 dkg Mehl.

Zusatzkarte 1 kg 96 dkg Brot oder 1 kg 40 dkg Mehl.

Verkauf nur nach Gewicht gegen Vorlegung der Ausweiskarte und Abtrennung der entsprechenden Abschnitte zulässig.

Nicht übertragbar! Sorgfältig aufbewahren! Nachdruck verboten!

Strafbestimmungen.

Zuwiderhandlungen werden an dem Verkäufer, wie an dem Inhaber der Karte mit Geldstrafen bis zu 5000 K oder mit Arrest bis zu 6 Monaten geahndet Bei einer Verurteilung kann auf den Verlust einer Gewerbeberechtigung erkannt werden. Fälschung der Ausweiskarte wird nach dem Strafgesetze bestraft.

FünfAbschnitte = 1 kleiner Laib (Wecken) Brot (35 dkg) oder ¼ kg Mehl.

Gemeindesiegel:

(coupon sections, repeated:) 70 g Brot oder 50 g Mehl

Figure 7.1 Ration Card for Bread in 1917

Source: Regional Archive Maribor.

grams of bread per week. Despite rationing, the supply of bread, rolls, and similar products was irregular and often depended on one's good luck and ingenuity; a personal acquaintance with a baker helped considerably. Written accounts from this period indicate that ration cards alone were insufficient and did not guarantee the desired foodstuffs.

In the early hours of the morning women gathered in front of bakeries and butcheries, in order to try and ensure at least some food for their families. According to one writer,

> Everyone who had been through this remembers those four hungry war years with a certain relief that they are now a thing of the past. What a wonderful feeling it is to be sated! And how terrible famine is! For most people wartime was a time of misery, sorrow, and trouble of all sorts. Yet by far the hardest was hunger... The young of today do not know the delight of savouring bread. They do not know how hard it is if there is none. The sombre present and the dark times that lie ahead are but a pale shadow of those past days shrouded in smoke, dripping with blood. Let us hope that those times will never return! The mothers and wives who started to queue in complete darkness outside bakeries and butcheries to obtain their share of the meticulously measured, pitiful food allotted to them by a piece of paper could tell you something about waiting hours for their turn. It was just the same as when a crowd of people is jostling in front of a train station ticket counter to get, for a quarter of the regular price, as many tickets as possible for a patriotic event. It seemed as if a multitude of people were trying to get into a pilgrimage church, but the house of God was too small for all of them.
>
> At home, children, the sick, and the elderly waited for the mother and the housewife to bring something to eat. Hungry and weary, they waited. Whatever would be placed in front of them on the table would have been insufficient even the previous day. Perhaps it would have sufficed the day before and would have prevented their strength from seeping away so visibly. Is it then strange that children withered and shrivelled, and the sick died unnecessarily?[4]

In 1916 the already difficult situation worsened: the quantity of flour per person was reduced, and sugar, coffee, lard, meat, potatoes, milk, tobacco, and clothes were rationed as well.[5] In addition, meatless and fatless days were introduced. In April 1916, the *Slovenec* newspaper reported on the effects of milk deficiency, particularly on children.

> This acute shortage of milk is affecting the entire Austrian territory, including our province. Milk is difficult to obtain, even for children. Food prices have increased so much that the less privileged classes can hardly afford the basic

---

4   Gaberc 1935: 174.
5   Mlakar 2004: 15.

daily foodstuffs anymore and are forced to substitute them with milk which, in comparison, is much cheaper. If milk, the most important and often the only affordable nourishment of the poor, becomes scarce the consequences will be considerable and may affect the entire population. There will be disease, and children will suffer stunted growth. Even today, it is already possible to see children in the street who are pale and sunken, which is the result of malnutrition. It is the duty of every farmer to feed his cattle as well as possible, ensuring adequate quantities of milk for the market. However, it is clearly difficult for farmers to deliver the substantial obligatory amount of hay to the army while being obliged to feed their milk cows with quality fodder. Our Provincial Board has therefore established a central body, which has already started to operate, that will provide farmers with high-quality animal feed. All farmers are therefore advised to feed their milk cows and draught animals with this feed that is highly nutritious and increases milk yield.[6]

Larger towns set up public kitchens providing food to those who had no family of their own or were not part of a larger household, for example single workers such as miners, but the kitchens did little to improve the existing situation. One veteran of the home front remembered that in his childhood the food supply depended mostly on the benevolence or otherwise of shopkeepers, bakers, and butchers. The possession of ration coupons did not in itself ensure that the holder would actually obtain the desired staples.[7] Since it became increasingly difficult to purchase certain foodstuffs, organized protests broke out from time to time. People were most affected by the shortage of flour, and in Maribor, for example, workers gathered to protest against the inadequately organized food supply and the lack of food, demanding from the district commissioner Dr. Krammer not only more flour, but also potatoes, lard and meat.[8]

In order to pacify the general unrest, those who were most at risk such as children up to the age of three were allowed access to meat at lower prices. Certain products, for example sausages, were occasionally distributed free of charge.[9] At times people were able to buy meat from horses too old for battle, which gave some relief to the hungry population. In addition, certain substitute foods were made available that prior to the war had been relatively unknown in Slovene food culture. Examples include cod, mass-produced pasta and kohlrabi as a replacement for potatoes. Smokers were particularly affected by the shortage of tobacco, which sometimes caused withdrawal symptoms. Since tobacco was available only through ration cards, and even then in meagre quantities at a high price, smokers often mixed it with dried walnut, vine, clematis, or turnip leaves.[10]

---

6   *Slovenec* 10 April 1916: 4.
7   Golija 1996: 34.
8   Golija 1996: 34.
9   Mlakar 2004: 15.
10 Mlakar 2004: 15.

In 1918, a citizen was entitled to only 750 grams of flour, 30 grams of lard, and 550 grams of potatoes weekly, and 750 grams of sugar per month. The situation deteriorated month by month. The most frequent victims of hunger were children, prompting approximately two hundred women to protest in the streets of Ljubljana and in front of the seat of the Carniolan provincial assembly. Foodstuffs were not only difficult to obtain but also very expensive, and wages were low. While the latter barely increased during wartime, the price of food increased by as much as 3,000 per cent, and according to the estimates of some historians, the general standard of living diminished fivefold.[11]

Life was no easier in the countryside. Fran Saleški Finžgar, a Slovene writer wrote:

> You don't have to tell me about war – the husband in Russia, the brother lying in a grave in Galicia, the horse and the wagon God knows where – and she, with her five children, has remained as the head of the family, a handyman, and a horse, all in one person.[12]

Since the men were fighting on the battlefields, in addition to taking care of the children, the sick, and the elderly, farm women were forced to do all the farm chores, a feat all the more difficult because they now had no draught animals. Nevertheless, work in the field had to be done, so women, aided by children and sometimes the elderly, had to undertake it themselves. At harvest time and on other occasions when more hands were essential, additional workers, mostly refugees and prisoners of war, were supplied by work agencies.

The Shortage of Basic Foodstuffs and their Substitutes

Since wartime brought a critical shortage of food for the civilian population both in towns and in rural Slovenia, this led to increasing hunger and vulnerability to disease, particularly in the second half of the war. The amount of rationed food decreased from year to year, and the irregular food supply made it uncertain whether one could obtain any at all. Only painstaking queuing in endless lines, luck or a personal acquaintance with shopkeepers, provided a small amount of bread, flour, meat and sugar. As a result, the black market flourished, and recipes were devised advising homemakers how to prepare savoury dishes from food substitutes.

Especially stressful was the lack of bread and flour. Provided that one presented his or her ration card coupon, the latter was obtained directly from a mill. Soon after the outbreak of war, it was forbidden to bake buns and other fancy breads. Bakers were allowed to bake bread only in loaves, but these could not be made with the best quality patent flour. Wheat flour had to be mixed with other types of

---

11   Himmelreich 1998: 6.
12   Finžgar 1965: 304.

flour and other ingredients, for instance pulses such as chickpeas, kidney beans, and fava beans, and also chestnuts and potatoes. Confectioners were not allowed to use more than 30 per cent wheat flour for their pastries.[13] These new bread products had an inferior taste and did not keep long. A commentator recalled:

> Regulations on what had to be added to flour for bread were increasingly strict; in reality, though, additives even included sawdust and all kinds of inedible stuff. Even warm bread did not have that beautiful fragrance of bread that is dreamed of by millions of hungry living beings, by millions of people regardless of where they live. After some time such bread acquired a weird taste and smell, and one's teeth could not continue biting into that tasteless mess. And yet... one wants to live! So we stood in lines in front of bakeries, patiently waiting, fingering our green ration cards in order to get a half loaf of such bread once or twice a week. The green ration cards for bread were a more coveted currency than the Austrian crown.[14]

Equally short of bread were soldiers on the battlefields, and many were already starving in 1915:

> A soldier had to endure double trouble, famine in addition to hardship. Soldiers fighting at the rear began to go hungry as early as 1915. Army bread was made of a terrible mixture of corn and kidney bean flour; sometimes it seemed that shredded and ground straw had been added to this mixture. It was so fragile that a loaf broke into pieces and crumbs if it fell on the floor. This kind of food was not filling at all, of course, especially since everything else that a soldier consumed was equally bad and of meagre quantity... In wartime everything except meat was a pitiful substitute for the real thing. All ersatz foods: chicory and acorn for coffee, strawberry, blackberry and other plant leaves for tea, and turnip and carrot peelings badly-washed, sometimes even mixed with sand, as vegetables.[15]

It is evident that ersatz bread could not satisfy the craving for real bread. One of the key substitutes for bread dough was that made with potato, which could be used to make the yeast itself as well as a substitute for wheat flour. Many recipes for the use of potatoes have been preserved: the yeast, which was prepared from boiled potatoes, leaven, water, and barley flour was generally used for making buckwheat, barley, and corn flour bread.[16]A shortage of coffee was overcome by preparing ersatz coffee, made from roasted ground barley, fava beans, lentils, and chicory. A coffee substitute could also be made from cumin, dried dandelion roots, grape seed,

---

13   *Straža* 20 August 1915: 5.
14   Reisman 1939: 55.
15   Gaberc 1935: 175.
16   Remec 1915: 127.

dried black figs, and acorns. These substitutes were blended with a small quantity of coffee beans and, valued as a treasure, drunk only occasionally at home.

Although prior to the war most people ate meat only periodically, and then only in modest quantities, they were forced to deal with a new level of scarcity. The already limited quantities had to be stretched even further by eating crackling, lard, and mixing lard with minced meat. Even these became increasingly hard to obtain, and at the start of the final year of the war, a ration card holder could buy no more than 30 grams of lard weekly. Since meat, meat products and lard were also very expensive, townspeople made use of the black market or smuggled food from the countryside into the towns. In Trbovlje, miners and their family members helped farmers with their farming chores and received food as payment. Some also traded certain goods for food, for example clothes, footwear, furnishings and other household items. This process of exchange, called the *berija*, had traditionally helped the mining community and other working-class families at earlier times of crisis and need.[17]

Townspeople throughout Slovenia tried to improve their diet by obtaining foodstuffs from their relatives and acquaintances living in the country, where food was more easily available. In spite of severe restrictions, the latter were sometimes able to conceal food to sell on the black market. Written sources mention crowds of people setting out for the country on Sundays, among them smartly dressed ladies who were not in the least embarrassed to haul heavy rucksacks crammed with food back to town.[18] This helped to alleviate the urgent shortage of foodstuffs in urban families where the lack of food was the most pressing. In rural areas, farmers still had some extra food, particularly flour, lard, meat and meat products that they had squirreled away to be used in times of need. A series of letters written by a farmer's wife to her husband on the front reveal how, left to fend for herself with seven underage children, she managed to work the farm by concealing, selling, and trading extra food for help with farm chores. In 1916, one letter stated:

> The animals are healthy and getting fatter. Pigs are a bit more expensive, but not as expensive as other foodstuffs. I have just sold 6 piglets, 4 to your brother and 2 to Micka living down the ravine. I worry how I will be able to hide the meat once people start stealing, and how I will work the land. We will have nothing to eat unless the soil is tilled.[19]

The situation on the battlefronts was no easier, serving in various units of the Austrian army, Slovene soldiers often reported to their families on the lack of food and the poor quality of the meals they received. By far the most difficult period was the last year of the war:

---

17 Mlakar 2004: 15.
18 Reisman 1939: 61.
19 Stanonik 1997: 27.

All the mulberries and cherries had been eaten as they were still ripening in the trees. As soon as wheat in the fields started to produce grain, it was stripped off by hungry soldiers. This took place after the very last dog and cat had vanished from the vicinity of Cessalto.[20]

The continuing hunger forced people to consume types of meat that had generally not been seen on Slovene dinner tables before the First World War. Civilians in towns were able to purchase the meat from old battle horses, and according to some sources, cats and dogs ended up being eaten by humans as well. Such meat was sometimes served in gravies or as meatloaf, usually with a number of condiments to conceal its origin. An ex-soldier from Slovenia who was fighting on the front in Northern Italy wrote:

> I have already mentioned that there were neither cats nor dogs in Cessalto and its vicinity. Dogs were freely killed by our soldiers and prepared as a supplement to their meagre soldier's ration. But cats stay closer to home, so it was often necessary to lure them to slaughter... A deputation of soldiers once went to a house where they had a pretty well-fed cat, considering the circumstances. They used gestures to make the housewife understand that the division general had an infestation of mice in his quarters. The woman thought the whole affair somewhat odd but lent them her cat. She did not see it again.[21]

A Wartime Cookbook

The shortage of food during the war forced Slovene homemakers to use foods that prior to the war had seldom, or indeed never, been consumed. There was increased interest in the preparation of fruit and plants growing in the wild, particularly mushrooms, dogwood and other berries, nettles, dandelion, horsetail, and strawberry and blackberry leaves. Advice was sought on how to prepare dishes from root crops such as turnips, cabbage and potatoes, and ideas were offered on how to use the various substitutes, potatoes, bread, and pasta leftovers.

Bread and pastries were made with corn, barley or buckwheat flour, none of which had generally been used in baking before the war. Since recipes for such dishes were scarce, the Katoliška Bukvarna Publishing Company published a cookbook with recipes that were considerably different from those in previous Slovene cookbooks. This collection of recipes for hungry times was titled *Varčna kuharica* [The Frugal Cookbook], and subtitled *Zbirka navodil za pripravo okusnih in tečnih jedi s skromnimi sredstvi: Za slabe in dobre čase sestavila Marija Remec* [A collection of recipes for the preparation of inexpensive, tasty and filling dishes for bad as well as good times: compiled by Marija Remec]. The

---

20   Gaberc 1935: 175.
21   Gaberc 1935: 176.

volume sold out surprisingly quickly and was reprinted after the war. Its preface, written by the author, states that the book provides for all the needs of a modest household:

> Thus it takes into consideration all the requirements of unpretentious cuisine – frugality as well as taste; we have strictly avoided giving instructions for lavish and wasteful dishes. Economy is the principal criterion for our husbands; good homemakers whose main concern and purpose are centred in the kitchen are invaluable. A frugal homemaker is a skilful one, combining all the noble characteristics of a conscientious wife and mother.[22]

While the frugality and carefulness of a homemaker were highly valued characteristics before the war, they became indispensable during wartime. The author particularly recommended the use of home-grown food rather than the preparation of expensive imports more suited, as she puts it, to 'lavish circumstances'. The cookbook's structure does not immediately suggest that it was created during the trying wartime period. Its numerous recipes are arranged in the usual general order: soups, thickeners, vegetables, gravy, meat dishes, salads, egg dishes, miscellaneous dishes, fish, leavened dough dishes, fancy breads, layer cakes, cookies and homemaking advice. However, a detailed analysis shows that most recipes are designed to be filling, using potato, corn, buckwheat, or barley flour, or a mixture of pulses and other foodstuffs. There are recipes that either had been unknown prior to the war or would have been used infrequently on particular occasions, for instance, kidney beans with apples or rice, noodles with potatoes, potato cake, plums with cornmeal, herring with barley and dumplings made from groats. Since cod, previously unknown in Slovene households, was among the foods that could be purchased with ration coupons, there were quite a few recipes for it, either as the main course or as an ingredient in other dishes. Besides vegetables, the section on side dishes recommends plants grown in the wild. There is a recipe for the preparation of boiled nettles, dandelion and sorrel.[23] Prepared in the same manner as spinach, the plants are first boiled, then chopped, seasoned and thickened with a roux. In an attempt to increase their popularity, Remec concludes by emphasizing the health benefits of such recipes.

The section on bread and fancy breads contains mostly recipes on the preparation of leavened dough made from corn, buckwheat or rye flour, and it particularly recommends the use of potato yeast. Only two recipes use white flour which was difficult to obtain during wartime; the first is for an inexpensive *potica* (nut roll) and the second for a traditional Czech Easter bread. Cake recipes clearly illustrate the difficult circumstances in which this cookbook was conceived. Provided one possessed a certain amount of ingenuity and knowledge, however, one could make a cake even in those troubled times. Although the recipes indicate that the best

---

22   Remec 1915: 6.
23   Remec 1915: 22.

cake ingredients, for example patent flour, eggs, butter, and cream, were difficult to obtain and had to be replaced with whatever substitutes were available. There are several recipes for the preparation of corn and buckwheat cake and also a buckwheat cake with chocolate. The final recipe in this chapter is for a wartime cake, which required only one egg and no white flour or fat:

> Put into a bowl 7 spoons of black coffee, 7 spoons of milk, 280 g of rye flour, 150 g of sugar, 1 teaspoon of cocoa, 1 egg, finely shredded lemon rind, cinnamon and other spices, and 1g of baking powder. Beat all the ingredients well, pour the batter into a cake pan and bake in a rather hot oven. When done, spread with jam and dust with sugar.[24]

Even though the structure and size of the volume, not to mention the variety of its recipes, do not explicitly state its origins, *The Frugal Cookbook* was written for use in wartime. At first glance it is no more modest than cookbooks published before the First World War; it is only close analysis of its recipes and the ingredients necessary for their preparation that indicate it was written during a period of severe food shortage and hunger. Its great popularity and the fact that it sold out so quickly indicate that homemakers during the war wanted instructions on how to prepare satisfying and relatively tasty meals; meals that in those grim times could, at least to some extent, bring joy to their families, even if that were only through the consumption of a mock roast or a wartime cake.

Conclusion

The First World War took not only the political elite by surprise, but also the rest of the population of Austria-Hungary who were unprepared for its length, course, cruelty and ultimately its conclusion. In anticipation of a brief and victorious war, the Austrian government had not planned for the actuality; this became obvious not only on the battlefront but also in both urban and rural civilian life. A chaotic food policy and inadequate food supply, which became an urgent problem from the spring of 1915 onwards, deleteriously affected the everyday life of the civilian population. The introduction of food ration cards neither alleviated the situation nor ended the shortage of food. The purchase of available foodstuffs, which were still in short supply and distributed sporadically, largely depended on luck and often on personal acquaintance with shopkeepers, butchers, and bakers. Civilians were forced to wait in endless lines for many hours to purchase their daily amount of war bread.

The consequences were terrible. Particularly in the third year of the war, famine was widespread throughout Slovene territory, with children and the elderly numbering the greatest proportion of its victims, but the exact number of civilian

---

24 Remec 1915: 176.

casualties of the First World War remains unknown. The authorities limited the food supply by introducing food ration cards, restricting the consumption of milk and sugar and by creating meatless days. In the final year of the war, the provisioning system collapsed completely, leaving the population to fend for themselves.[25] As a result, black marketing and profiteering flourished, as did food smuggling and bartering of food for other necessities of life. Hunger led to personal crises and the collapse of moral values, which drove people to commit acts they would never even have contemplated in peaceful times; Vinko Gaberc wrote in his 1935 memoir:

> All those who felt hunger, or at least observed it away from the front, know about its consequences... So many things could be obtained for bread. Somebody exchanged a loaf of bread for an official signature that would be hard to get in different circumstances; another swapped it with a townsman for a shabby suit; the third unscrupulously bent the will of a hungry woman. Many a virtue grew feeble in the face of famine. Who could be strong when hungry? I often went hungry. I guess I had not been full for three years... The same was true for many others; still, we managed to survive.[26]

References

Cvirn, J., Vidic, M., and Brenk, L. 2000. *Ilustrirana zgodovina Slovencev.* [Illustrated History of the Slovenes] Ljubljana: Mladinska knjiga.

Fridl, J., Kladnik, D., Orožen Adamič, M., and Perko, D., eds 1998. *Geografski atlas Slovenije.* [Geographical Atlas of Slovenia] Ljubljana: DZS.

Finžgar, F. 1965. *Prerokovanja* [Prophesies]. Celje: Mohorjeva družba.

Gaberc, V. 1935. *Brez slave – spomini na svetovno vojno.* [Without Glory – Memories of the First World War] Ljubljana: Publisher unknown.

Godina Golija, M. 1996. *Prehrana v Mariboru v dvajsetih in tridesetih letih 20. stoletja.* [Food Culture in Maribor in the 1920s and 1930s] Maribor: Založba Obzorja.

Himmelreich, B. 1998. *Prva svetovna vojna v dokumentih Zgodovinskega arhiva Celje.* [The First World War in Documents from the Regional Archive Celje] Celje: Zgodovinski arhiv.

Mlakar Adamič, J. 2004. *Teknilo nam je!* [It was to our Taste!] Trbovlje: Zasavski muzej.

Reisman, A., 1939. *Iz življenja med vojno.* [About Life during the War] Maribor: publisher unknown.

Remec, M., 1915. *Varčna kuharica.* [The Frugal Cookbook] Ljubljana: Katoliška bukvarna.

---

25 Svoljšak 2004: 156.
26 Gaberc 1935: 181.

Svoljšak, P. 1993. Prva svetovna vojna in Slovenci, [The First World War and the Slovenes] in *Zgodovinski časopis* [Historical Review], 47:2, 263–87.

Svoljšak, P. 2004. Tudi jaz sem pomagala do velike zmage, [I too, have helped win the great victory] in *Ženske skozi zgodovino* [Women through History] edited by A. Žižek. Ljubljana: Zveza zgodovinskih društev Slovenije.

Slovenec newspaper, 1916.

Straža newspaper, 1915.

Stanonik, M. 1997, *Štiri matere-ena ljubezen: Zgodba neke družine.* [Four Mothers – One Love: The Story of One Family] Ljubljana: Slovensko etnološko društvo.

Chapter 8

The Spanish Civil War and its Aftermath: Eating Strategies and Social Change[1]

Alicia Guidonet Riera

Introduction

Food is a very complex phenomenon, determined by various social, cultural and historical factors. Analysis of its role allows us to understand some of the human relationships which enable social reproduction and which take place within this specific context. Furthermore, the study of food crises, such as that considered here, is becoming increasingly significant. As suggested by Mauss, the strengths and weaknesses of social cohesion are areas that are subject to particular analysis during times of crisis:

> A good way to analyze the strength or weakness of the foundations of social cohesion is to examine those times when it disappears... Moments of panic, departures for war, vendettas, battle manœuvres, collective rages and frenzies, mass exodus, mystic wanderings, collective extasis, fears caused by calamities and epidemics – all these are nothing but variations of the same reality, whose causes and effects are equally important. This reality often means the death of those supra-organic conglomerates that are groups and sub-groups. In the most extreme situations this may mean the disintegration of society, sometimes even its complete disappearance.[2]

It also seems timely to remember that the diachronic study of various critical situations related to food shows that this phenomenon may vary widely, depending on the historical, social and cultural nuances of the context studied.[3] Furthermore,

1   This chapter presents some of the results of more extensive research undertaken in Catalonia in 2006 and 2007. The main objective of the research was to recover oral memory concerning the food of people who had lived through the Spanish Civil War and postwar period. 33 narratives were collected, and these were complemented with audiovisual and bibliographic information. The project was financed by the Centre for Promotion of Catalan Popular and Traditional Culture (CLT/349/2006), the Agency for Management of University and Research Grants (AREM 2005) and the Caixa de Sabadell.
2   Mauss 1969: 328.
3   Ferrières 2002.

these periods of food crisis are not always related to a lack of food. By way of an example, there is one type of food crisis, found in modern societies, which arises in situations of plenty and is connected to the consequences of industrialization and globalization: the paradigmatic case is that of 'mad cow disease'.[4] With that in mind, we now turn to the focus of study that concerns us here, which involves a lack of food, to a greater or lesser extent, in the context of the Spanish Civil War and the postwar period (1936–55). Our subjects are a group of men and women who lived through this period, and who had to undertake everyday tasks relating to food, production and/or supply, processing, distribution, consumption and disposal, under conditions that were not favourable and in some cases, desperate.

The great wealth of information we have obtained obliges us to define the objectives of this study. We will focus on an analysis of some of the means of food supply that appeared during this period. Another look at Mauss' study mentioned above raises the following questions: what happens when social crisis comes about as a consequence of a war, and how does this affect food habits? What strategies do people use to survive? Apart from the material aspects of food, what role does it play in the reproduction of a social group's values?

The First Survival Strategy: Diversification of Resources

In a context in which food is scarce or lacking, acquiring it becomes an urgent priority. This particular aspect can be seen very clearly in the interviews carried out, and the information provided by the people we worked with focuses almost exclusively on this area. It can therefore be concluded that during the period of time studied, obtaining food in order to survive became one of people's main concerns; indeed their memories are fundamentally focused on the strategies they used in order to achieve this end.

While obtaining food is a basic task that takes up much of the time of those enduring a situation of crisis due to a lack of food, optimizing all types of resources becomes a key strategy for these people. A diversification of supply channels takes place; this happens at various levels, including obtaining the money to purchase food by doing various tasks, use of the informal economy, and accessing the assistance networks that were available.

The increase in supply channels in times of crisis due to a lack of food has been demonstrated by various authors, who despite working in different contexts, all agree that this hypothesis holds true. González Turmo mentions overcoming hunger and the lack of food in rural areas by means of 'poor man's bread', i.e. food that was easy to obtain because it is part of the natural environment, this included hunting small game, river fishing and picking fruit.[5] The study by Aguirre analyzed the practices of the urban poor in Argentina during five crisis periods: the years

---

4   Guidonet 2010.
5   González Turmo 2002: 301.

1989, 1991, 1995, 1997 and 2001. The author makes a distinction between the four sources of food income: urban labour markets (formal and informal), social welfare, mutual assistance and self-supply.[6] Perianu, in the context of the final decade of Communist power in Romania, talks about the 'subsistence networks' that made it possible to obtain food and which operated in parallel with the formal channels. These networks existed thanks to a wide variety of interpersonal relationships: friendship, acquaintance, employment or pure exchange.[7] Finally, Plancade, in his work on the food eaten by the inhabitants of a shack on the outskirts of a large city, mentions the different ways they obtained food in order to be able to eat. These included doing odd jobs to earn money, such as guarding cars, donations of money or food thrown away by nearby restaurants and the exchange of food, or food and services, with other residents of the neighbourhood.[8]

Our study makes a distinction between the different supply strategies in the context of the Spanish Civil War and the postwar period including formal and informal work, self-sufficiency, exchange and welfare networks. However, given the vast number of nuances in each of these channels, in this study we will focus our attention on analysing reciprocity, which is a specific type of exchange.[9] Based on this concept, we will see that eating, far from being a merely physiological function, is also used to construct an individual's identity, and indeed, to promote the reproduction of the values of a given social group.

Acquiring Food by Social Morality: Generalized Reciprocity and Balanced Reciprocity

The study by Marcel Mauss, *The Gift: Forms of Function and Exchange in Archaic Societies*, describes the role of the gift as an action involving one of the following transfer mechanisms: giving/receiving or returning/receiving.[10] This symmetry, or reciprocity, can be explained by taking different variables into account. As noted by various authors, reciprocity can thus be defined as a moral standard, which 'structures the giving and the return of help'.[11] It is a standard with boundaries that may change depending on the following factors: the equivalence of value, the time transpired between the gift and it being reciprocated and the social distance involved.[12] Sahlins uses these factors to structure three types of reciprocity, as mentioned by Narotzky.[13]

---

6   Aguirre 2005.
7   Perianu 2008.
8   Plancade 2008.
9   Narotzky 2004:71.
10 Mauss 1968.
11 Narotzky 2004: 73.
12 Narotzky 2004: 73.
13 Sahlins 1977, Narotzky 2004: 73–6.

The first is generalized reciprocity. This is characterized by the lack of equivalence of a stipulated value, a minimal social distance, such as between family members, friends or neighbours and no clear expectation of a time of return. What predominates is, on the one hand, the recipient's need, and on the other, the donor's expectation that they will receive help if this becomes necessary in the future. Our study shows the use of relationship networks, such as neighbours, friends, work colleagues or an extensive family network, in order to obtain food. In her study, Aguirre gives a very precise definition of generalized reciprocity when she explains that social networks act as a 'social security system' which 'channels messages, goods and services from those who have most to families in a critical situation, who in turn return the favours received to their friends, neighbours and relatives when they are in need'.[14] According to the author, the people who receive this help therefore do not only obtain food, but also the implicit message of care and protection from those helping them, i.e. of safety in a crisis situation. At the same time, they reinforce and or maintain existing social links: between friends, neighbours and/or family members. Most importantly, they help the social reproduction of the group. Our study contains numerous narratives in which the help received from those closest to the recipient is apparent. This help has distinctions depending on whether the gift is food or an invitation to eat:

> We had relatives in Sant Climent, who came to sell things in the square, and so they gave us things as well ... potatoes, fruit, a bit of fruit as well, apples, anything, and that was good for us ... as they were relatives, they gave us things...
>
> My aunt Carmen took me to Cal Foix, which was a baker's in Sarrià, and she knew the baker, who gave me little pieces of stale bread...
>
> When I really ate well was when I went to my fiancé's house, because there his father worked, so did my fiancé and his sister, there were three wages coming in and they ate well there..

Other examples provide a very clear illustration of some of the characteristics defining generalized reciprocity, such as the fact that the time taken to return the help is not precisely defined, or that equivalence of value is not stipulated in these cases. This can be seen in two examples. In the first, we consider a family consisting of a father, mother, son and daughter, whose circumstances changed when the war broke out. They took in a relative from their extended family network, the grandfather, who arrived from Tremp, a village in the province of Lleida, in order to seek help from the family. This hospitality was given in exchange for small chores performed around the home, such as looking after the vegetable garden 'our grandfather came from Tremp and came to our house ... grandfather and I looked after the vegetable garden and my mother looked after the animals...'

The second example shows the help received by a widow and her daughter from their extended family network, that is, the mother's sister. The women

---

14   Aguirre 2005: 125.

had just arrived from exile and obtained some of the food they ate from their relative. The person telling the story, the daughter, makes it very clear that the help received was returned years later: 'My aunt gave us a lot of food, which she sent to us from Girona. That helped us very much ... my mother paid her back later, but I mean, at that time, it saved us...' Interestingly, this generalized reciprocity exceeded some of its characteristics in the context studied. This was the case with social distance, which according to Sahlins' description, should be 'minimal'. But this was often not the case, as the distance defined was much greater than the family, the neighbours or the local community. In these cases, it can be seen how the conceptual void that arises is replaced by other factors, such as ideology and beyond that, the desire to perpetuate class values. In short, the aim was to reproduce a social group that was in danger of disappearing, or being reduced or silenced, due to the war. This was the case with gifts exchanged between those with ideological affinities. In agreement with other authors, such as Plancade, this shows that the symbolic value of food in a crisis in which food is lacking fulfils a very important role, one that extends beyond mere nourishment. We will look at two examples given by women, girls at the time, who crossed the frontier over the Pyrenees when the Nationalist troops occupied Catalonia:

> When we reached the frontier they gave us a glass, I seem to remember that it was either milk or hot chocolate...
>
> My most vivid memory of food is when we were going towards Brittany, the train stopped at stations and people came out and gave us sandwiches ... they did it out of solidarity, they didn't know us ... it was out of solidarity...

Balanced reciprocity is different from generalized reciprocity in that it arises within a less dense social fabric, in which the rules of how to behave are much clearer. In other words, the time transpiring between the gift and its reciprocal gift is ideally immediate, and the exchange of goods is identical.[15] The narratives we gathered include many situations which show the existence of this type of exchange. This data enables us to analyze first the value of the goods exchanged, and second, the levels of exchange according to the social status of the people involved in the transaction.

If we look at these narratives, the exchange of goods takes place when there is a surplus of a product and there is an attempt to exchange it for another product that is lacking or for one which is urgently needed, for example, when there is a child or someone is ill. That is why many of the exchanges take place between the country and the town, or between agricultural areas specializing in the cultivation of just one foodstuff, such as rice in the Valencia region. According to Moreno, the distribution of rations was more efficient in the towns, meaning that it is not surprising that some of these rationed products, such as sugar and soap, were much sought after by the peasants:

---

15   Narotzky 2004: 75.

People exchanged things, some people brought eggs and exchanged them for something else they didn't have, for oil, for example, everything had its own value ... I used to barter things but during the war we had soap and we went to the peasants who didn't have it and exchanged it for potatoes and came back with a big sack of potatoes weighing 20 kg. We had to go to Masnou, we had to go up the stream, it was all fields, we had soap because of rationing ... and if maybe you put two or three rations together, you had more soap than you needed and they didn't have soap ... that happened at the end of the war...[16]

As mentioned above, the level of the exchanges was not the same for everyone. The stories told to us provide some tools for the consideration of this exchange value, which varied according to the purchasing power of the individuals participating in the transaction. One example we collected shows how a man, the owner of a wine cellar, exchanged wine for goods like meat, clothing and even furniture: 'We were able to exchange wine for chicken drumsticks, clothes, furniture ... everything was based on barter...that lasted all through the war.' Goods, such as wine and furniture, were not commonly used as mediums of barter. More usually it was basic food products that were exchanged such as almonds, carobs, sweet potatoes and corn.

Obtaining Food Following the Breakdown of Moral Standards: Negative Reciprocity

Theft, stealing, swindling and haggling frequently appear in the informants' testimonies as strategies to obtain food. This resulted from a 'breakdown, transformation or suspension of the moral order'.[17] This issue is critical. If 'positive reciprocity', i.e. 'giving in order to receive' as described by Mauss was based on a shared morality, then theft, stealing, swindling and haggling are part of a different category, defined as 'negative reciprocity', based on the suspension of the accepted moral order. In their studies of reciprocity in Nazi concentration and extermination camps, Moreno and Narotzky identify the need to make a distinction between the two categories of reciprocity.[18] Negative reciprocity is part of a model of '(take in order to) give/(ask for in order to) receive/(keep in order to) be'. It would enable survival in times of crisis such as periods when food is scarce.

As mentioned above, this subject frequently appears in the testimonies gathered. It is evident that the suspension of moral standards results in a disordered situation that leads to varied and differing reactions. It can be seen how this factor leads to conflicts, either of a personal nature that affect the individual who has the opportunity to steal, swindle or haggle, or which arise among individuals who feel that they have been victims of these actions. This can be seen in various

---

16 Moreno 2005: 152.
17 Moreno and Narotzky 2000: 127.
18 Ibid., Narotzky and Moreno 2002.

cases. The first shows two very different reactions to the opportunity to break into three abandoned train wagons full of food. According to the testimony of the witnesses, in the winter of 1939 during the days after the fall of Barcelona to the Nationalist forces, chaos reigned in the city. A number of people took advantage of the situation in order to obtain food from the numerous stores and hiding places where the Republican army had kept its provisions. This 'taking' of food after breaking into the stores and hiding places was seen by some as theft, and in some cases a successful break-in and the consequent removal of food was prevented. For example, one woman, who was 16 years old at the time the incident took place, told us of her personal dilemma during such a situation and the argument she had with her sister, whose opinion was diametrically opposed. While our witness felt unable to break into the wagon of a train and take food, her sister rebuked her, taking advantage of the occasion to obtain some food:

> A neighbour came to our house and said to us "Go to the station, everyone's breaking in to the wagons of the train..." As we were so hungry, my sister said, "I'll go and take a look..." and I followed her. When I saw what was happening, I burst into tears! I started crying when I saw all the wagons full of people picking up sacks. I couldn't bring myself to go there. My sister could. She said "Well, if someone else is going to have it, I'll have it myself', and she picked up some lentils and some beans... and I was crying... "That's stealing! That's stealing!"... my sister said "Can't you see we're hungry? Haven't you been hungry enough? Don't you see that if you don't take it, someone else will?"

On a more general scale, the use of this type of strategy was accepted. It was justified as something that helped people to survive difficult times, despite the fact that some witnesses made a moral judgement of such actions:

> When the war started things were very bad, my father started working as a chef, he worked for those who were in the war because ... there wasn't any coal... [he was a coal merchant] ... he worked as a cook again, here in Barcelona, the Hotel Victoria was where the Corte Inglés department store is now, and there was a big store on the ground floor with a lot of shops, it was called Vicenç Ferrer, it still exists, but now it's on Carrer Comerç, and the Hotel Victoria was above it. There came a point where the chickens were so thin we couldn't serve them, we took no pleasure in killing them and eating them, because they were so thin that they were falling over because they were hungry as well. Well, my father made lentils or rice or something like that, he was working in the Hotel and the Hotel Glaciar which was on the Ramblas and you came out through Plaça Reial – now it's a shop, or in the Hotel Espanya, where my father had worked as well. He used to make me come out of school and go and see him because they always gave me some biscuits, my father's workmates, or they used to give me some cans of Russian meat that were like cold meat, and he gave me the husks of the lentils, what was left over, for the chickens to eat, but underneath it all he used to put a

tin of Russian meat, it was hidden ... it had a jelly, a packet sort of meat, it was stringy meat as if it was a beef stew ... they said it came from Russia. Cracked or spoiled rice, they used to boil it and mash it up and the leftovers of the rice, ... my father kept it for the chickens ... because my father was always highly regarded for getting food for us, really he was stealing ... it was stealing, but he did it for us, looking at it objectively, he was swindling them...

The social reproduction of the group also takes on different meanings depending on who is involved in the action concerned. At this point, we must mention the *estraperlista*, or black marketeer. Originally, *straperlo* was the name of a game of chance, similar to roulette, that could be rigged. Attempts were made to introduce it into Spain during the years before the war broke out.[19] It is likely that because of its connection to the idea of getting rich illegally, the word ended up being used to refer to the black market as a means for supplying food during the most difficult times in the conflict and in the postwar period. The dictionary defines the word as 'illegal trading in products confiscated by the state, sold at exploitative prices'.

The Franco regime's policy of self-sufficiency was responsible for the appearance of the black market. The cutback in imports concentrated more power in the hands of just a few people, who had close links to the regime. This led to an increase in corruption and made it easy to amass great riches by using this parallel market. Black market trading also took place on a small scale, among weaker social groups that often had no alternative given the desperation of their situation.[20] The testimonies obtained make a distinction between these two types of black market, as well as between two contradictory opinions. The black marketeers who got rich during this period were strongly criticized by those on the defeated side:

If you had cash you could get hold of more food ... obviously, eh?, Because then there was the black market ... Then came the black market time, but on the quiet ... and they were all Nationalists, weren't they? The ones involved in the black market, the ones doing that, were on the Nationalist side, and we were the ones buying! They were the ones that were selling! A bunch of spivs, they were! There was one of them, whose name was Ramonet, he was a black marketeer, he made money left right and centre, another one who stayed ... you know that little palace on the Passeig de Gracia, if you go right to the top ... the Palau Robert, well that belonged to one of them, one of those Catalans who came in with the Nationalists, he was a black marketeer! He made a packet, loads of money! And he had all sorts of women ... They sold the food through middlemen, didn't they? They didn't do it directly ... they sold it for higher prices ... you had to pay them more... "if you want to eat I can get it for you," in the market because there were people there ... they talked to people – "listen, if you want meat or if you want something" ... that was the middlemen, you see? I can get it for you at such and

19   Guillamet 1995: 30–38.
20   Barranquero and Prieto 2003: 61.

such a price, "OK", because people were trying hard to get hold of food ... they arranged to meet you at a stand ... we hardly ever had to do that ...

However, as mentioned above, this strategy was not only used by people with links to the Franco regime in order to get rich, but also on a small scale by people trying to survive in the midst of poverty and a lack of food. In a study carried out in the province of Malaga, Barranquero and Prieto highlighted the large number of women involved in this other type of black market, called the 'domestic' black market.[21] This had a great deal to do with these channels being used by women who, as a consequence of the war, had become their family's sole breadwinner. The following example is particularly illustrative. The mother of the person who told us this story had to separate from her partner when the war ended, as he was forced into exile in France. He never came back. The narrator's mother, who at that time lived in Torroella de Montgrí, a village in Catalonia, became involved in black market trading, and managed to survive and feed her daughter:

> My father left in February 1937 and I stayed here with my grandfather and my mother. My mother had worked in a textile factory but of course, that all went ... and as she was left alone there, in the war she worked making material for the war, but afterwards it was very difficult ... and she got involved in the black market and I remember she used to go to the farms looking for corn, and she brought it to me, and I remember when we used to come back from the farms at night or from Ullà, and she sold it in paper wraps, half a kilo, a kilo, to people who had chickens ... mother used to go and sell to the farms with a basket: she sold bran – I didn't go with her, though, did I? I only went at nights to keep her company ... She took soap to the farms that she made herself with oil and fat because she also sold black market olive oil, and went around the houses. When they made oil the remains were always left over and they sold that and mother bought it to make soap, or instead sometimes the oil was rancid, because of course, they must have put lots of things in it so that it went further ... She also sold a lot of beans from the remains of sorting, because of course, beans are sorted, aren't they? So in Palafrugell, which was bigger and more industrial ... they had nothing to eat, and so they came here. We divided the beans into three, we sorted them in the evening and there were three categories...

It is obvious that the breakdown in social norms in a given context is tolerated to a greater or lesser extent depending on various factors, such as the perceived need of the person involved, whether it is for existence or abused for profit. We have also seen how despite the situation of need, breaking the rules can sometimes be almost unbearable. This was illustrated by the example of the girl who did not want to take food from the stores that had been broken into during the chaotic period of the fall of Catalonia to the Nationalists.

---

21   Barranquero and Prieto 2003: 226–7.

A final example suggests the reasons that may have lain behind a refusal to infringe the established norms in a critical situation. It also reinforces the hypothesis that eating, even at times when food is lacking, contains significant aspects, which make eating – or not doing so – a communicative action that reinforces group identity. It also maintains and passes on class values. That is what happened to the person whose experience is described below. This woman had been forced into exile in France with her parents; conditions were very difficult and her food consisted in the main of items that had been picked or gathered from fields and hedgerows. The local school children had enough food and there were even leftovers, which generally ended up in the dustbin.[22] The woman explains that she had no problems with picking fruit on the way to school from home, but never picked up leftover food that the French boys and girls had thrown away.[23]

> If they had any food left over, the children threw it into a type of bin in the yard, and I remember that I walked up to that bin many times, but I never actually picked up a piece ... and I was only seven years old ... I never did get to try that food because my mother talked to me about responsibility, freedom, and dignity ... and I suppose that I was passing those values on ... I suppose that out of dignity, in my own way I thought that I couldn't pick up that food that the French children had thrown away...

Conclusion

An analysis of food during the Spanish Civil War and the postwar period shows the importance of reciprocity during this period. Various social agents used this reciprocity for social reproduction. According to Narotzky and Moreno, the concept of social reproduction involves various aspects: the material, in how to reproduce the things necessary for survival; political, in the reproduction of power relationships; cultural, in terms of the reproduction of identity; and moral, with regard to the reproduction of a just system.[24]

Having analysed the reciprocity associated with food during a time of scarcity, it is possible to establish the existence of each of these aspects. Obtaining food was a key priority, given the basic need for food in order to survive. Despite this, for

---

22   This case involves a situation on the boundary of negative reciprocity, because something that has been thrown away does not strictly belong to anyone. However, the strongly symbolic nature of the event is obvious. The boys and girls had just thrown their food away and they are still in the place where the event occurs, which means that, at a symbolic level, the food is still part of the children.

23   This action would also have entailed symbolic appropriation of the values of the French boys and girls, that in the geographical area studied, were clearly opposed to those of the Spanish Republicans.

24   Narotzky and Moreno 2002: 300.

example, we have seen how black marketeers could reproduce and maintain power relationships between the various social groups. Defence of the local community, of the household, neighbours and friends is also apparent, involving attempts to retain social identity and encourage its survival during a period of crisis. In specific terms, reproducing specific values may, as shown above, lead to a breakdown in the prevailing norms. For example, this happened when a witness refused to eat food despite being hungry. This would have entailed losing the dignity upheld by those who had been defeated, who would fight for decades, despite being in exile, for the value system which was extinguished by their defeat. In conclusion, eating – or not doing so – is closely related to relationships of power, identity and morality. It should be emphasized that these relationships go far beyond purely material aspects, which are also extremely important at times when food is lacking.

References

Aguirre, P. 2005. *Estrategias de Consumo: qué Comen los Argentinos que Comen.* [Strategies of Consumption: What Argentineans Eat] BuenosAires: Centro Interdisciplinario para el Estudio de Políticas Públicas.

Barranquero, E. and Prieto, L. 2003. *Así Sobrevivimos al Hambre: Estrategias de Supervivencia de las Mujeres en la Postguerra Española.* [How We Coped with Hunger: Survival Strategies of the Women of Spain in the Postwar Period] Málaga: CEDMA.

Ferrières, M. 2002. *Histoire des Peurs Alimentaires: Du Moyen Âge à l'Aube du XXe Siècle.* [The History of Food Scares from the Middle Ages to the Twentieth Century] Paris: Seuil.

González Turmo, I. 2002. Comida de Pobre, Pobre Comida, [Food for the Poor, Poor Food] in *Somos lo que Comemos. Estudios de Alimentación y Cultura en España* [We are What We Eat: Spanish Food and Cultural Studies], edited by M. Gracia. Barcelona: Ariel.

Guidonet, A. 2010. ¿Miedo a Comer? Crisis Alimentarias en Contextos de Abundancia [Afraid to Eat? Food Crisis in Times of Abundance]. Barcelona: Icaria.

Guillamet, J. 1995. *Tots Hem Fet l'Estraperlo.* [We Have Used the Black Market] Barcelona: Columna.

Mauss, M. 1969. *Cohésion Sociale et Divisions de la Sociologie.* [Social Cohesion and Divisions of Sociology] Euvres T-3. París: Minuit.

Mauss, M. 1968. Essai sur le Don. Formes et Raison de l'Echange dans les Sociétés Primitives [The Gift: The Form and Reason for Exchange in Archaic Societies] Paris: PUF.

Moreno, R. 2005. Pobreza y Supervivencia en un País en Reconstrucción, [Poverty and survival during the reconstruction of a nation], in *Pobreza, Marginación, Delincuencia y Políticas Sociales bajo el Franquismo* [Poverty, Exclusion,

Crime and Social Politics under Franco's Dictatorship], edited by C. Mir et al. Lleida: Universitat de Lleida.

Moreno, P. and Narotzky, S. 2000. La Reciprocidad Olvidada: Reciprocidad Negativa, Moralidad y Reproducción Social. [Reciprocity's dark side: negative reciprocity, morality and social reproduction] *Hispania*, LX(1), 127–60.

Narotzky, S. 2004. *Antropología Económica. Nuevas Tendencias* [New Directions in Economic Anthropology] Barcelona: Melusina.

Narotzky, S. and Moreno, P. 2002. Reciprocity's dark side. Negative reciprocity, morality and social reproduction. *Anthropological Theory*, 2(3), 281–305.

Perianu, C. 2008. Précarité Alimentaire, Austérité [Food insecurity and austerity. Eating in the last decade of communism in Romania] in *Anthropology of Food* [Online], 6; available at: http://aof.revues.org/index4513.html [accessed: 15 November 2008].

Plancade, A. 2008. 'Les Aliments des Habitants de la 'Cabane' [Food in the shack: potluck and leftovers' in *Anthropology of Food* [Online], 6; available at: http://aof.revues.org/index4672.html [accessed: 15 November 2008].

Sahlins, M. 1977. Economía de la Edad de Piedra. [Stone Age Economics] Madrid: Akal.

Chapter 9

Alimentary and Pellagra Psychoses in Besieged Leningrad

Pavel Vasilyev

Introduction

This chapter explores mental health implications of famine such as alimentary and pellagra psychoses in the context of besieged Leningrad in the Second World War. The period under investigation is not limited to the siege (1941–44) but extends into the postwar years until the early 1950s. This is due to the fact that particular attention needs to be paid to the longer-term impact of malnutrition on mental health. Another important feature of this chapter is that alimentary and pellagra psychoses are perceived broadly, that is as medical, social and scientific phenomena.

The Second World War presented a major challenge to food supply and distribution throughout Europe, but the most terrible cases of malnutrition and famine were arguably witnessed in the Soviet Union. Confronted with the loss of the most important agricultural territories early in the war,[1] the Soviet authorities had to establish a rationing system and food cards for the civilian population.[2] Moreover, there was a deliberate imbalance in food distribution. As William Moskoff put it, '[Soviet] institutions worked better to feed the army than they did to feed the civilian population'.[3] Of course, malnutrition also affected the military (the battle of Stalingrad serving as the 'metaphor for hunger in the Red Army'),[4] but the worst conditions were experienced by civilians especially in blockaded Leningrad. The siege of Leningrad, which lasted from September 1941 to January 1944, was one of the crucial military operations on the Eastern front during the Second World War. It has received a lot of attention from scholars, but further study is relevant and essential as contemporary Russian historians re-evaluate the legacy of the war and particularly its social and cultural aspects.[5]

The siege of Leningrad was an unparalleled phenomenon in European and global food history. John D. Barber called it an 'extreme demographic

---

1   Moskoff 2002: 2.
2   Voronina 2009: 35.
3   Moskoff 2002: 2.
4   Ibid.
5   See e.g., Seniavskaia 1995, Livshin and Orlov 2003, and Sevost'ianov 2004 vol. 2.

catastrophe ... [and] unique event in the history of mankind'.[6] One of the most lethal and longest military sieges in modern history, it caused large-scale famine. Hunger was primary. It was the most significant factor that determined the whole history of Leningrad during the war.[7] William Moskoff has strikingly described the desperate supply conditions that persisted in the city throughout the siege and especially in the 'hungry winter' of 1941–42.[8] Ration norms were clearly inadequate – reaching 250 grams of bread for the workers and 125 grams for other categories of the population at the lowest point in November 1941.[9] The social and health implications of the Leningrad famine are among the most important – though relatively unexplored – topics.

This chapter focuses on little-known consequences of malnutrition such as alimentary and pellagra psychoses. A number of historians have mentioned this topic, albeit briefly, and it is necessary to provide a short historiographical review. The Soviet period witnessed the publication of several classic books on the history of Leningrad during the war, which also examine medical and demographic subjects.[10] The best-known early Western account was written by Harrison Salisbury, who discussed many questions that were then considered taboo in the Soviet Union.[11] Nikita Lomagin, one of the most famous contemporary historians of the Leningrad siege, has applied many innovative approaches in his research.[12] There is also an extensive literature on specific problems such as public health,[13] food supply[14] and science[15] in the besieged city. Several works use oral history which provides us with horrific narratives of the siege.[16] Of special importance is famous and tremendously wide-ranging *Blokadnaia kniga* [The Blockade Book].[17]

Investigation of alimentary and pellagra psychoses in besieged Leningrad is impossible without using primary sources. This chapter draws on Soviet medical texts written during the war or in the late 1940s.[18] These texts provide us with first-hand experience of the doctors and they also demonstrate that most research on the topic was conducted in this period. Archives emerge as another useful source of

---

6   Barber and Dzeniskevich 2001: 5.

7   Ibid.: 5.

8   Moskoff 2002: 185–96, 199–203.

9   Avvakumov 1944: 119.

10   Azarov et al. 1967, Koval'chuk et al. 1985.

11   Salisbury 1969.

12   Lomagin 2002.

13   Brozek, Wells and Keys 1946, Gladkikh 1985, 2006, Magaeva 2001, Barber and Dzeniskevich 2001, 2005.

14   Moskoff 2002.

15   Kol'tsov 1962, Sobolev 1965, 1966.

16   Dickinson 1995, Loskutova 2006. For an excellent example of oral history in regard to food shortages, see Chapter 8 by Alicia Guidonet Riera.

17   Adamovich and Granin 1994.

18   Gel'shtein 1943, 1947, Chernorutskii 1947, Shmar'ian 1948, Golant and Miasishchev 1948, Abramovich 1949.

information about everyday medical practice in the besieged city as well as research activities in the postwar period.[19] Some scientific publications and research reports used in this chapter have not been examined by scholars before. These materials are of special interest, because, as Andrei R. Dzeniskevich puts it, they 'were for internal use only'[20] and the texts of subsequent conference presentations and publications were greatly modified. The official documents related to the medical aspects of the siege of Leningrad are of twofold quality. On the one hand, there is a considerable amount of information compared with other cases of large-scale famine. This is not surprising because Leningrad was a developed city with a modern health care system in the 1940s. On the other, the terrible conditions of the siege and growing despair among many officials led to the fact that many records are imprecise and misleading.[21]

The main purpose of this chapter is to study the influence of malnutrition on mental health of the population of the city of Leningrad during the military siege and in the early postwar years. The first section describes the development of scientific understanding of alimentary and pellagra psychoses throughout the period. Subsequent sections classify mental violations caused by nutritional dystrophy and avitaminosis and discuss the influence of psychoses on everyday life in the besieged city. Finally, this chapter looks at relapses and the longer-term impact of alimentary and pellagra psychoses in the postwar period.

Overcoming Scientific Challenges: Medical Research on Alimentary and Pellagra Psychoses

It is extremely important to mention that nutritional dystrophy in general was quite a new medical problem for Soviet physicians. Even more so, alimentary and pellagra *psychoses* were almost totally unexplored topics. Some research on the impact of long-term malnutrition on mental health was conducted during the First World War (Bonhoeffer, and, especially, Stifler on the siege of Przemyśl). Soviet

---

19   Among relevant archives are: *Tsentral'nyi gosudarstvennyi arkhiv nauchno-tekhnicheskoi dokumentatsii Sankt-Peterburga* [Central State Archive of Scientific and Technical Documentation of St. Petersburg; hereinafter referred to as TSGANTD SPb] (funds of V. M. Bekhterev Psychoneurological Research Institute and Leningrad branch of All-Union Institute of Experimental Medicine), *Arkhiv Muzeia V. M. Bekhtereva Sankt-Peterburgskogo nauchno-issledovatel'skogo psikhonevrologicheskogo instituta im. V. M. Bekhtereva* [Archive of V. M. Bekhterev Museum of V. M. Bekhterev St. Petersburg Psychoneurological Research Institute; hereinafter referred to as AMB], *Tsentral'nyi gosudarstvennyi arkhiv Sankt-Peterburga* [Central State Archive of St. Petersburg] (funds of Leningrad City Soviet and S. M. Kirov Leningrad State Institute of Postgraduate Education for Physicians), and *Sankt-Peterburgskii filial Arkhiva RAN* [St. Petersburg branch of the Archive of the Russian Academy of Sciences].

20   Barber and Dzeniskevich 2001: 98.

21   Ibid.: 14–15.

psychiatry also paid some attention to the neurological and mental disorders caused by starvation (mostly Bekhterev,[22] and later Iudin, Rozenshtein). Nevertheless, these diseases remained largely unknown to the majority of physicians.[23] In fact, the first detailed academic descriptions of alimentary and pellagra psychoses were published in the Soviet Union during the Second World War.[24] This lack of knowledge was aggravated by serious flaws in the organization of medical research that were inherent in the prewar Soviet health care system.[25]

However, the outbreak of the war stimulated research activity as a whole,[26] and the deteriorating conditions in Leningrad forced doctors to concentrate on new diseases that were emerging in the besieged city.[27] During the siege the staff of the Leningrad branch of All-Union Institute of Experimental Medicine conducted complex research into nutritional dystrophy, work that is often considered to have been of extreme importance.[28] Among other effects, Professor Nina P. Kochneva studied changes in the functioning of both central and peripheral nervous systems as well as mental disorders that were caused by the decreasing percentage of carbohydrates in patients' blood.[29]

The results of this research were shared in professional conferences that were quite regularly held even in the terrible conditions of the siege. In 1941–43, the staff of State Institute of Postgraduate Education for Physicians organized 22 conferences which included more than 60 lectures.[30] The conclusions of the scientific conferences that were held at V. M. Bekhterev Psychoneurological Research Institute in 1942–43 are accurately reflected in the documents of Central State Archive of Scientific and Technical Documentation of St. Petersburg.[31] A conference in January 1942 was specifically devoted to the problems of neuropsychiatry.[32] In April 1943, the decision was made to organize special scientific committees attached to the Department of Public Health of Leningrad City Soviet's Executive Committee. The specific task of these research groups was to study alimentary dystrophy and avitaminosis.[33] The research by Leningrad physicians conducted during the war was acclaimed by both the professional

---

22   See Vserossiiskoe soveshchanie po voprosam psikhiatrii i nevrologii 1919.

23   Sobolev 1965: 17. See also TSGANTD SPb, f. 313, op. 2–3, d. 9, l. 7, 33.

24   Sobolev 1966: 71.

25   Ibid.: 22.

26   Ibid.: 8. About the importance of research activity during wartime, see TSGANTD SPb, f. 313, op. 2–3, d. 9, l. 2, and Sobolev 1966: 3–6.

27   Barber and Dzeniskevich 2001: 3.

28   Ibid.: 106.

29   Sobolev 1966: 71.

30   Sobolev 1965: 18.

31   TSGANTD SPb, f. 313, op. 1–1, d. 219–a.

32   Sobolev 1966: 71.

33   Sobolev 1965: 17.

community and the Soviet authorities.[34] In the spring of 1942, physicians organized short-term courses on nutritional dystrophy and avitaminosis in all the districts of Leningrad to inform a wider audience about the research results.[35] However, the subsequent deterioration of welfare and health care systems presented a serious challenge to medical practice.[36]

After the end of the war, nutritional dystrophy and its various implications remained among the most popular research topics. Some scholars rated the scope and quality of this scientific work highly.[37] The most important papers on alimentary and pellagra psychoses written during this period are, 'Psychoses Under Nutritional Dystrophy and Vitamin Deficiencies' (1948),[38] and 'Mental Disorders During Military Siege of Leningrad' (ca. 1950),[39] by Raisa I.A. Golant and Vladimir N. Miasishchev. D. I. Rozenshtein's 'Neurological and Mental Disorders Related to Nutritional Dystrophy'(ca. 1950),[40] concentrated on psychoses caused by siege pellagra.[41] Nevertheless, it is important to stress that alimentary and pellagra psychoses were still not very prominent on the research agenda of corresponding institutes.

One of the most mysterious problems related to alimentary and pellagra psychoses in besieged Leningrad is the termination of investigations on this topic in the early 1950s. Subsequently, no traces of relevant research activity can be found in archival documents.[42] This is even stranger if we compare it to the situation in the early postwar years when several high-quality works were published. There are several possible explanations for this change, but each has some flaws.

First, it can be argued that the total number of patients with alimentary and pellagra psychoses was relatively small.[43] However, this could be countered by the fact that a great number of people died without receiving any medical treatment due to the enormous problems in everyday health care practice. Secondly, as B.E. Maksimov has argued about depressive disorders, 'the siege greatly increased the number and, even more important, the acuteness of mental disorders'.[44] A detailed study of living conditions in the besieged city may even lead us to question whether there was anyone in blockaded Leningrad who was 100 per cent mentally sound according to the criteria set by official Soviet medical science.

---

34   Ibid.: 17. For positive evaluation of that work, see also TSGANTD SPb, f. 313, op. 2–3, d. 9, l. 2.

35   Sobolev 1965: 17.

36   Barber and Dzeniskevich 2001: 13.

37   Ibid.: 3.

38   TSGANTD SPb, f. 313, op. 2–3, d. 9.

39   TSGANTD SPb, f. 313, op. 2–3, d. 11.

40   TSGANTD SPb, f. 313, op. 1–1, d. 485, l. 7–9.

41   TSGANTD SPb, f. 313, op. 2–3, d. 25.

42   See, *e.g.*, TSGANTD SPb, f. 313, op. 1–2, d. 10, 11, 277, 278.

43   Barber and Dzeniskevich 2001: 122.

44   TSGANTD SPb, f. 313, op. 2–1, d. 182, l. 28. See also Barber and Dzeniskevich 2001: 121.

Another explanation rests upon the nature of psychiatric patients. The authorities might not have considered them that important and some scholars have shown that psychiatric hospitals and relevant research groups were rather poorly financed.[45]

The closure of investigations on alimentary and pellagra psychoses may also be attributed to censorship. Psychoses can be seen as the extreme version of health problems that were caused by siege malnutrition. Thus, the re-orientation of research activities in the early 1950s may be linked to the Leningrad affair[46] and the subsequent re-evaluation of the role of the city during the Second World War. However, this hypothesis does not explain why some changes can be traced back to 1946, when the Leningrad affair was not yet being planned.[47] Finally, the shift in research plans can be related to the changes in international relations. Andrei R. Dzeniskevich argues that the intensifying Cold War forced medical research institutes to concentrate primarily on the impact of radiation on human organism.[48] Thus, less topical concerns such as alimentary psychoses were no longer included in research plans. Nevertheless, the documents of V.M. Bekhterev Leningrad Psychoneurological Research Institute do not entirely correspond with this argument.[49]

From Slow Thinking to Morphological Brain Changes: The Mental Consequences of Starvation

An analysis of works of Soviet psychiatrists allows us to describe and classify mental disturbances that were caused by nutritional dystrophy and avitaminosis. It is important to mention that there were many negative factors besides malnutrition in besieged Leningrad which contributed to the rise of mental disorders. Among them were the cold, a deterioration of welfare and health care systems, permanent bombing, death of friends and relatives, decreasing capacity for work, common

---

45   Barber and Dzeniskevich 2001: 121.

46   The Leningrad Affair (*Leningradskoie delo*) refers to several fabricated cases that were brought against Soviet party functionaries in the late 1940s and early 1950s. The affair was not confined to Leningrad *per se*, but most of the persecuted party members originated from Leningrad, worked there during the period of the siege and became very popular. The most famous victims of the Leningrad Affair were Nikolai Voznesenskii, Aleksei Kuznetsov and Petr Popkov, all of whom were executed on 1 October 1950.

47   Barber and Dzeniskevich 2001: 105.

48   Ibid.: 106. Compare also John D. Barber's view of physicians as pragmatics who concentrated largely on urgent practical issues of health care, Barber and Dzeniskevich 2001: 3.

49   TSGANTD SPb, f. 313, op. 1–2, d. 10, 11, 277, 278.

apathy and depressive mood, and general uncertainty.[50] Vladimir N. Miasishchev considered these additional factors to be of extreme importance.[51]

Alimentary and pellagra psychoses caused slow and hampered thinking along with physical and psychic adynamia[52] that often resulted in morphological brain changes.[53] Grigorii B. Abramovich (school of Raisa I.A. Golant) specifically emphasized the long duration of the diseases, rich development of depersonalization experiences and general variety of manifestations.[54] Another important problem was the dynamics of the psychoses.[55] Vladimir N. Miasishchev attributed the rise of alimentary psychoses to events in January 1942. While somatic symptoms of nutritional dystrophy disappeared in the early summer of 1942 due to relative improvement of nutrition, alimentary psychoses persisted well until 1944. Avitaminotic (pellagra and to a lesser degree scurvy) psychoses appeared somewhat later, in the spring of 1942, and increased dramatically during this year.[56] Doctors repeatedly noticed this gap between the appearance of somatic and mental manifestations of malnutrition. To some extent, this was advantageous and it also provided an opportunity to make the separation between alimentary and pellagra psychoses clearer.[57]

The treatment of alimentary and pellagra psychoses was a considerable challenge for Leningrad physicians because all possible options were seriously hampered. One of the prerequisites was normal nutrition, which was extremely difficult to implement. Other forms of treatment included vitamin B$_1$, amphetamine, neostigmine, and metamizole.[58] The siege experience also showed that psychic activity and general 'energetic condition' were crucial.[59]

'People Gone Mad With Hunger': New Perspectives on the Social History of Besieged Leningrad

The focus on alimentary and pellagra psychoses in besieged Leningrad also helps us to open up new avenues of research. Case studies recorded by Soviet physicians make it possible to reconstruct features of the extraordinary everyday life in

---

50   TSGANTD SPb, f. 313, op. 2–3, d. 9, l. 1–2; Barber and Dzeniskevich 2001: 13.
51   Miasishchev 1948: 21, TSGANTD SPb, f. 313, op. 2–1, d. 231, l. 15, Barber and Dzeniskevich 2001: 122–3. See also AMB, f. VII, d. 5, l. 2.
52   Smirnov 1951: 193, TSGANTD SPb, f. 313, op. 2–3, d. 9, l. 7.
53   TSGANTD SPb, f. 313, op. 2–1, d. 231, l. 10.
54   AMB, f. VII, d. 5, l. 2.
55   Ibid.
56   Barber and Dzeniskevich 2001: 122. See also: TSGANTD SPb, f. 313, op. 2–3, d. 9, l. 2–3.
57   TSGANTD SPb, f. 313, op. 2–3, d. 9, l. 2–3.
58   Sobolev 1966: 71.
59   TSGANTD SPb, f. 313, op. 2–1, d. 231, l. 15, Barber and Dzeniskevich 2001: 122–3.

the besieged city. These show the health status of the population of Leningrad at a micro level. Thus, medical histories emerge as a new source on the social history of Leningrad during the siege. This includes topics such as subjective perceptions of the world and its impact on everyday life. This material is relevant not only with regard to documented patients of psychiatric hospitals, but also the broader community. A substantial part of the symptoms of alimentary and pellagra psychoses could be observed in a great number of patients with borderline personality disorders. Moreover, many Leningrad citizens died without receiving any medical treatment or having been diagnosed with a mental disease.[60] The outbreak of the war was itself a major stress for many people, but it is useful to remember that it was preceded by the Great Purge and the Winter War with Finland. The overall atmosphere of terror that persisted in Leningrad in the early 1940s makes it possible to assume that a substantial part of the population was predisposed to alimentary and pellagra psychoses.

The psychoses changed basic elements of perception and social interaction such as time, space, speech, wishes and motivation. Typical features included constant expectation of revelation,[61] expanding space[62] and time,[63] and, more characteristically, total social apathy and indifference.[64] However, the most interesting among specific mental disturbances were a morbid concentration on the physical sensation of hunger and related hallucinations. These were reflected both in popular culture and in medical texts.

In survivors' accounts, phrases like 'people gone mad with hunger' or 'hunger made people crazy' are abundant.[65] Obsession with food was often explicitly linked to cannibalism. In the medical context, Raisa I.A. Golant's work gives many examples of mental patients roaming the city who saw and heard hallucinations related to food ('they invited me to treat me well', 'table covered with various foods and beverages'). This concentration on the feeling of hunger led them to disregard all cultural norms and they became indiscriminate in choosing means to attain their ends. Thus, these patients often engaged in antisocial and criminal behaviour – such as malicious prosecution, reporting on their friends and neighbors to the authorities, eating rats alive or stealing foodstuffs and other items.[66]

---

60   For the account on the practical problems of psychiatry in Leningrad in the 1940s, see, *e.g.*, Barber and Dzeniskevich 2001: 13.

61   Adamovich and Granin 1994: 197.

62   Ibid.: 366.

63   TSGANTD SPb, f. 313, op. 2–3, d. 9, l. 22.

64   Survivors of the siege often recollect that at some point their consciousness 'became dull'. They were no longer afraid of anything (including bombings and hunger) and felt absolutely indifferent about their own and their relatives' possible deaths, Loskutova 2006: 114, 115.

65   Moskoff 2002: 197. Cf. Dickinson 1995, Loskutova 2006: 115–16.

66   TSGANTD SPb, f. 313, op. 2–3, d. 9, l. 8, 12–14, 17, 29, and TSGANTD SPb, f. 313, op. 2–3, d. 11, l. 6–7.

The legacy of the siege affected mental health which could give rise to alimentary and pellagra psychoses in the postwar period.[67] It is crucial to emphasize the possibility of relapses (due to both physical and mental factors) which persisted for a long time after the siege was lifted and food conditions improved. First, it is important to mention that even those people who managed to survive the siege of Leningrad and were not diagnosed with a psychiatric disease, could encounter mental disorder or depressive conditions later in their lives.[68] Further, patients who were hospitalized with alimentary and pellagra psychoses during the siege frequently experienced relapses in subsequent years.[69] Contributing factors were deteriorating food and living conditions, stress, physical strain and injuries.[70]

Conclusion

This chapter has analysed alimentary and pellagra psychoses in besieged Leningrad as complex phenomena, whose medical and scientific understanding varied over time. Importantly, Soviet medical texts demonstrate the development of doctors' perception of the physiological consequences of starvation. Doctors' attempts to diagnose and classify mental breakdown caused by starvation further provide a new perspective in the effort to reconstruct features of the social history and the history of everyday life in the besieged city.

The prospect of further research on alimentary and pellagra psychoses in besieged Leningrad includes analysis of a broader range of primary sources and also hermeneutics of medical texts. Among other interesting topics are the impact of psychoses on social interaction and a comparative history of alimentary and pellagra psychoses. The necessity of research on the impact of malnutrition on the mental health of Leningraders should be underlined, too.

References

Abramovich, G.B. 1949. *Pozdnie psikhozy na pochve alimentarnogo istoshcheniia: (Klinika i psikhopatologiia).* [Tardive Psychoses Caused by Nutritional Dystrophy: (Clinic and Psychopatology)] Leningrad: Goslestekhizdat.

Adamovich, A.M. and Granin, D.A. 1994. *Blokadnaia kniga.* [The Blockade Book] St. Petersburg: Pechatnyi Dvor.

Avvakumov, S.I. ed. 1944. *Leningrad v Velikoi Otechestvennoi voine Sovetskogo Soiuza: Sbornik dokumentov i materialov.* [Leningrad in the Great Patriotic

---

67   For the importance of such research, see Barber and Dzeniskevich 2001: 3–4, 265.
68   Loskutova 2006: 114–15, 117.
69   TSGANTD SPb, f. 313, op. 2–3, d. 9, l. 11, 13, 17.
70   TSGANTD SPb, f. 313, op. 2–3, d. 9, l. 11–12, 30.

War of the Soviet Union: Collection of Documents and Materials] vol. 1. Leningrad: Gospolitizdat.

Azarov, V.B. et al. 1967. *Ocherki istorii Leningrada.* [Studies in the History of Leningrad] vol. 5. Leningrad: Nauka.

Barber, J.D. and Dzeniskevich, A.R. eds 2001. *Zhizn' i smert' v blokirovannom Leningrade: Istoriko-meditsinskii aspect.* [Life and Death in Besieged Leningrad: Medical Historical Aspects] St. Petersburg: Dmitrii Bulanin.

Barber, J.D. and Dzeniskevich, A.R. eds 2005. *Life and Death in Besieged Leningrad, 1941–44.* Basingstoke: Palgrave Macmillan.

Brozek, J., Wells, S., and Keys, A. 1946. Medical aspects of semi-starvation in Leningrad (siege 1941–1942). *American Review of Soviet Medicine,* 4(1), 70–86.

Cherepenina, N.IU. ed. 2002. *TSentral'nyi gosudarstvennyi arkhiv Sankt-Peterburga: Putevoditel.* [Central State Archive of St. Petersburg: A Guidebook] 2 vols. Moscow: Zven'ia.

Chernorutskii, M.V. ed. 1947. *Alimentarnaia distrofiia v blokirovannom Leningrade.* [Nutritional Dystrophy in Besieged Leningrad] Leningrad: Medgiz.

Dickinson, J.A. 1995. Building the blockade: New truths in survival narratives from Leningrad. *Anthropology of East Europe Review,* 13(2), 19–23.

Gel'shtein, E.M. 1943. *Metodicheskie ukazaniia po raspoznavaniiu i lecheniiu alimentarnogo istoshcheniia.* [Methodical Instructions on Diagnosing and Treating Nutritional Dystrophy] Leningrad: N.p.

Gel'shtein, E.M. 1947. *Alimentarnaia distrofiia.* [Nutritional Dystrophy] Moscow: Medgiz.

Gladkikh, P.F. 1985. *Zdravookhranenie blokadnogo Leningrada, 1941–1944 gg.* [Public Health in Besieged Leningrad, 1941–1944] 2nd Edition. Leningrad: Meditsina.

Gladkikh, P.F. 2006. *Zdravookhranenie i voennaia meditsina v bitve za Leningrad glazami istorika i ochevidtsev: 1941–1944 gg.* [Public Health and Battlefield Medicine During the Siege of Leningrad Through the Eyes of Historian and Eyewitnesses] St. Petersburg: Dmitrii Bulanin.

Golant, R. IA. and Miasishchev, V.N. eds 1948. *Nervnye i psikhicheskie zabolevaniia v usloviiakh voennogo vremeni.* [Neurological and Mental Disorders Under Wartime Conditions] Leningrad: N.p.

Kol'tsov, A.V. 1962. *Uchenye Leningrada v gody blokady (1941–1943).* [Leningrad Scientists in the Years of the Siege] Moscow: AN SSSR.

Koniukhova, T.S. et al. eds 1997. *Kratkii spravochnik po fondam Tsentral'nogo gosudarstvennogo arkhiva nauchno-tekhnicheskoi dokumentatsii Sankt-Peterburga.* [A Short Guide to the Funds of Central State Archive of Scientific and Technical Documentation of St. Petersburg] St. Petersburg: Liki Rossii.

Koval'chuk, V.M. et al. eds 1985. *Nepokorennyi Leningrad: Kratkii ocherk istorii goroda v period Velikoi Otechestvennoi Voiny.* [Unconquered Leningrad: A Brief Outline of the History of the City During the Great Patriotic War] Leningrad: Nauka.

Livshin, A. IA. and Orlov, I. B. eds 2003. *Sovetskaia povsednevnost' i massovoe soznanie, 1939–1945.* [Soviet Everyday Life and Mass Consciousness, 1939–1945] Moscow: ROSSPEN.

Lomagin, N.A. 2004. *Neizvestnaia blokada.* [The Unknown Siege] 2 vols. 2nd Edition. St. Petersburg: Neva.

Loskutova, M.V. ed. 2006. *Pamiat' o blokade: Svidetel'stva ochevidtsev i istoricheskoe soznanie obshchestva.* [Remembrance of the Siege: Eye-witnesses' Testimonies and Historical Consciousness of the Society] Moscow: Novoe izdatel'stvo.

Magaeva, S.V. 2001. *Leningradskaia blokada: Psikhosomaticheskie aspekty.* [The Siege of Leningrad: Psychosomatic Aspects] St. Petersburg: ABS.

Miasishchev, V.N. ed. 1948. *Nauchno-issledovatel'skii psikhonevrologicheskii institut im. V. M. Bekhtereva. Kratkaia istoriia Instituta (K 40-letiiu so dnia osnovaniia).* [V. M. Bekhterev Psychoneurological Research Institute. A Brief History of the Institute (Fortieth Anniversary Edition)] Leningrad: N.p.

Moskoff, W. 2002. *The Bread of Affliction: The Food Supply in the USSR During World War II.* 2nd Edition. Cambridge: Cambridge University Press.

Salisbury, H.E. 1969. *The 900 Days: The Siege of Leningrad.* New York: Harper & Row.

Seniavskaia, E.S. 1995. *1941–1945: Frontovoe pokolenie: Istoriko-psikhologicheskoe issledovanie.* [The Front Generation: A Study in Historical Psychology] Moscow: Institut rossiiskoi istorii.

Sevost'ianov, G.V. ed. 2004. *Voina i obshchestvo, 1941–1945.* [War and Society] 2 vols. Moscow: Nauka.

Shmar'ian, A.S. ed. 1948. *Nervnye i psikhicheskie zabolevaniia voennogo vremeni.* [Wartime Neurological and Mental Disorders] Moscow: Medgiz.

Smirnov, E.I. ed. 1951. *Opyt sovetskoi meditsiny v Velikoi Otechestvennoi voine 1941–1945 gg.* [The Experience of Soviet Medicine in the Great Patriotic War] vol. 27–8. Moscow: Medgiz.

Sobolev, G.L. 1965. Uchenye Leningrada v gody blokady, 1941–1943. [Leningrad scientists in the years of the siege] *Istoricheskie Zapiski*, 75, 3–25.

Sobolev, G.L. 1966. *Uchenye Leningrada v gody Velikoi Otechestvennoi voiny, 1941–1945.* [Leningrad Scientists in the Years of the Great Patriotic War] Moscow: Nauka.

Voronina, T. 2009. From Soviet cuisine to Kremlin diet: changes in consumption and lifestyle in twentieth-century Russia, in ICREFH X, 33–44.

Vserossiiskoe soveshchanie po voprosam psikhiatrii i nevrologii. [All-Russian Conference on Psychiatry and Neurology] 1919. Moscow: Narkomzdrav.

PART III
Home Front: The State Intervenes

Chapter 10

Fair Shares? The Limits of Food Policy in Britain during the Second World War

Ina Zweiniger-Bargielowska

Introduction

Food policy in Britain during the Second World War, characterized by extensive rationing and 'fair shares', is generally considered a great success which helped to maintain civilian health and morale in a 'people's war'. Fair shares and equality of sacrifice were prominent themes in wartime propaganda which fostered social solidarity in a long and often tedious war.[1] This interpretation has been questioned in recent literature which draws attention to persistent inequality, social conflict, discontent and the fragility of national unity on the home front.[2] My book, *Austerity in Britain: Rationing, Controls, and Consumption, 1939–1955*, contributes to this revisionist perspective by qualifying the myth of the home front characterized by egalitarianism and common purpose.[3]

This chapter builds on this research and it highlights the shortcomings of food policy during the war. The British flat-rate rationing system made insufficient allowances for differential needs and food shortages hit some social groups harder than others. Sacrifice was not equally shared between classes or between men and women and official distribution channels were bypassed in the black market. 'Fair shares' was a compelling slogan, but at times there were extensive doubts whether scarce supplies were really shared equally. Discontent about flat-rate rations was high among male manual workers, food policy demanded a disproportionate sacrifice from women and the wealthier social groups had greater opportunities to augment their diet with unrationed foods. The chapter further draws attention to the gap between policy, propaganda and public response. Welfare foods were targeted at children and pregnant or nursing women, but take-up was low and brown bread remained unpopular. Vital statistics generally registered improvements, but nutritional status was never monitored effectively.

---

1   For a summary of the literature see Addison 2005, Mackay 2002.
2   Fielding et al. 1995, Smith 1996, Rose 2003.
3   Zweiniger-Bargielowska 2000.

Food Policy and the British Diet during the Second World War

Wartime food policy was determined not just by nutritional requirements, but also by economic criteria such as anti-inflationary policy and the reduction of imports in view of the shipping shortage.[4] The system was based on flat-rate rations of protein foods, fats and sugar, complemented by unlimited supplies of bread, potatoes and restaurant or canteen meals. This so-called buffer was intended to satisfy differential energy requirements and communal feeding expanded considerably during the war.[5] Building on the experience of rationing in the First World War, a 'cardinal' principle was that the ration book should be 'as fully capable of being regarded as a guarantee' as it had been in the earlier war.[6] Distribution of supplies was based on a consumer–retailer tie and each consumer registered with a particular retailer who received rationed food to cover registrations. The Ministry of Food (MF) was established shortly after the outbreak of war and from January 1940 onwards rationing was gradually extended to bacon and ham, meat, cheese, fats, sugar, preserves and tea. These accounted for about one-third of energy intake and the bulk of animal protein, fats and sugar. Milk and eggs were subject to distribution schemes, which involved no definite entitlement and allowed for additional purchases depending on availability. Under the National Milk Scheme supplies were channeled towards children and pregnant or nursing women, who were also entitled to supplementary rations and welfare foods. In view of limited and seasonal availability, vegetables, fruit, fish and poultry were never rationed and perceived inequalities in supply, high prices and under-the-counter sales of these foods became a focus of discontent. One important innovation was the introduction of points rationing in December 1941. This new scheme, which covered processed foods including canned meat, fruit and vegetables, biscuits, cereal products, dried fruit and pulses, involved no registration. Instead consumers could spend their points at any retailer and the MF was able to manage demand by adjusting points levels. Boosted by lend-lease imports, the scheme was a great success and it was followed by the introduction of personal points for chocolate and sweets in the summer of 1942.

The wartime diet in Britain was relatively more plentiful than that in other belligerent countries. This was in part due to lend-lease supplies which accounted for ten per cent of energy and more than one-sixth of animal protein and fats in 1943 0–44.[7] After an initial reduction early in the war the food situation stabilized and an adequate energy and nutrient intake was maintained although consumption

---

4   For a short overview, see Oddy 2003: 133–66, Burnett 1979: 322–32. The three volume official history provides a full account, Hammond 1951, 1956, 1962, see also Hammond 1954.

5   See Chapter 11 by Peter Atkins.

6   The National Archives, Kew, (TNA), MAF 72/598, Rationing, 4 January 1937.

7   Ministry of Food 1946: 10–11.

of animal protein was 'considerably' lower.[8] The diet was characterized by increased consumption of compulsory brown bread, potatoes and milk. Simultaneously, there was a reduction of fats, and particularly butter, meat and bacon, sugar as well as fish, fresh fruit and eggs. This marked a reversal of interwar dietary trends. Processed and manufactured foods which had become staples were increasingly scarce and food quality was frequently poorer.[9] This virtual peasant diet was nutritionally adequate and healthy but also dull and monotonous. In order to understand the significance of wartime food policy in the history of the British diet it is necessary to look beyond averages. The combination of rationing, food subsidies and full employment put an end to the profound inequalities between income groups which had received extensive attention in the 1930s and no social group fell short of basic requirements. The policy 'revolutionized the social class distribution of the diet', this transformation persisted beyond the end of rationing and 'income-group differences in diet were never as great as they had been' before the war.[10]

Flat-rate Rationing and Inequality of Sacrifice

A flat-rate rationing system was easier to administer, but its equity is debatable because the system took insufficient account of diverse needs, unequal distribution within the household and differential access to scarce unrationed foods. This was acknowledged in a MF memorandum which noted that, 'Rationing is essentially inequitable; it provides the same quantity of an article for each person without any consideration of their needs or habits or of their capacity to secure alternatives'.[11] The policy 'bore most heavily' on people living alone and 'least' on families 'whose capacity for mutual adjustment was greatest'. It was assumed that single people generally had more money to spend and the policy was generous to families with young children, but rather less so with regard to adolescents. Access to restaurant and canteen meals or the possibility to grow vegetables and keep livestock could augment rations considerably. To some extent these 'inequalities and privileges ... cancelled each other out' and, for example, urban populations benefited from the former, whereas people in rural areas were more able take advantage of the latter.[12] There was considerable anxiety about food early in the war, but food worries became negligible as the supply situation stabilized and from 1942 onwards food rationing was perceived as the 'finest achievement in the war'.[13] This did not mean that these inequalities were irrelevant and the following paragraphs explore male manual workers' discontent with the inadequacy of rations, the burden borne by

---

8   Ibid.: 48–9.
9   Ibid.: 160–61, Oddy 2003: 160–62.
10 Nelson 1993: 116.
11 Quoted in Hammond 1951: 125.
12 Hammond 1954: 233–4.
13 TNA, INF 1/292, 30 June – 7 July 1942.

housewives who were primarily responsible for implementing food policy and the persistence of class differences in access to food.

According to surveys conducted in 1942–43, nearly a third of male workers in heavy industry were dissatisfied with food rationing.[14] A more detailed enquiry revealed that 42 per cent thought that they were not getting enough food to keep fit. This was attributed above all to insufficient meat, mentioned by over half, but also fats, bacon and sugar and 72 per cent thought that rations were not enough. The next most dissatisfied group was men in light industry, whereas a majority of women workers and housewives felt that they had enough food to keep fit.[15] This situation deteriorated subsequently and in February 1943 half of male manual workers considered their diet insufficient to keep fit, with another 11 per cent doubtful.[16] The possibility of an extra meat ration for heavy workers had been considered by a War Cabinet sub-committee but this was rejected because the Trades Union Congress was 'strongly opposed to any differentiation between heavy workers and other classes of the community' in view of the 'problem of identifying such workers'.[17] These grievances were addressed by the expansion of canteens and there was a special cheese ration for workers who could not be catered for in canteens. With plentiful bread and potatoes, discontent was not due to the lack of food but rather the shortage of highly prized items such as meat. Associated with strength and virility, meat was a traditional marker of status and male privilege in the working class diet.[18] Under rationing, working-class meat consumption was lower than prewar levels. This was widely resented at a time of full employment and Ernest Bevin, the Minister of Labour, maintained that the reduction of the meat ration in 1941 was a 'serious danger' to morale.[19]

Likewise, women's and particularly housewives' position in wartime food policy has to be located within a wider cultural context. The obverse of men's privileged access to more desirable and expensive items was female self-sacrifice and many working-class women lived on a diet dominated by bread and tea. This unequal distribution of resources in the household, which has been extensively documented in the early twentieth century, was generally accepted in working-class

14   Nuffield College Library, Oxford, (NCL), Wartime Social Survey, Food I: Food Schemes: A Collection of Short Reports on Inquiries made by the Regional Organization of the Wartime Social Survey, May 1942–January 1943.

15   NCL, Wartime Social Survey, An Inquiry into a Typical Day's Meals and Attitudes to Wartime Food in Selected Groups of the English Working Population, n.s. 16 and 19, 1942–43; based on inquires conducted between April and July 1942.

16   NCL, Wartime Social Survey, An Inquiry into (i) A Day's Meals, and (ii) Attitudes to Wartime Food in Selected Groups of British Workers, n.s. 32, June 1943; based on a survey conducted in February 1943.

17   TNA, CAB 75/27, War Cabinet: Home Policy Committee, Sub-Committee on Rationing, 21 October 1939.

18   Twigg 1983: 21–5, Oren 1973: 109–11, Ross 1993: 3–34.

19   TNA, CAB 67/9, WP (G) (41) 51, 21 May 1941.

culture because the health of the breadwinner had to be maintained.[20] A survey of working-class women in the 1930s suggests that this practice persisted and a third were found to be in poor health.[21] Rations were allocated on an individual basis, but consumption occurred at the household level and it was impossible to ensure that women actually consumed their full ration. There is no quantitative data, but it is likely that many wives and mothers gave some of their ration to hungry men and children. During the war housewives strongly supported rationing. They accepted sacrifice as their contribution to the war effort and generally considered themselves to be well fed. The survey speculated that this may be due to low expectations and when asked about the adequacy of their children's diet, women were rather more critical.[22] According to Hammond, feeding adolescents was a 'recurrent nightmare' particularly during school holidays.[23]

As feminist scholars have noted, the unequal division of resources in the household was not confined to material goods but also included factors such as time.[24] Women's tendency to spend their time caring for their families rather than indulging in personal leisure pursuits was utilized by the state as an indispensable aspect of food policy. Housewives were responsible for maintaining the health and morale of the civilian population with considerably reduced resources and their daily battle on the kitchen front played a critical role in the wider war effort. According to a MF leaflet, the 'line of Food Defence runs through all our homes ... A little saving here and there ... becomes an immense amount. ... *The woman with the basket* has a vital part to play in home defence. By saving food you may be saving lives'.[25] An avalanche of propaganda informed housewives of the details of food policy, advised on how to make the most of scarce resources and suggested new recipes such as 'mock' dishes. With the introduction of clothes rationing, soap rationing and shortages of virtually all household goods, housewives learned how to 'make do and mend' to maintain at least a semblance of customary standards and domestic rituals. This involved queuing for unrationed foods and other scarce items. Food queues were described as a 'bigger menace to public morale than several serious German air raids' in February 1941.[26] Queuing was mostly done by housewives, nearly a third queued in June 1942 and the proportion of women queuing was highest in the lower income groups.[27] Oren's conclusion, albeit based on an earlier period, that working-class women's 'elastic standard of living served as a buffer' in the household and the wider economy acquired a renewed significance.[28] Housewives

20 Oddy 2003: 60–62, 70.
21 Spring Rice 1939.
22 See note 15.
23 Hammond 1954: 233.
24 Seymour 1992.
25 Ministry of Food, *Wise Housekeeping in War-time*, n.d. [approx. 1940]
26 TNA, INF 1/292, 12–19 February 1941.
27 TNA, RG 23/7, 18.
28 Oren 1973: 121.

shielded men and children from the full impact of the reduction in consumption, a disproportionate sacrifice which extended well beyond the poorest income groups in the Second World War.

Class differences in access to food were reduced, but these were by no means eliminated. Restaurant meals were never rationed. For example, the diaries of Harold Nicolson and Sir Henry Channon contain numerous references to luncheons and dinners at hotels, restaurants and clubs.[29] According to morale reports in 1942 many people felt that 'everything is not fair and equal and that therefore our sacrifices are not worthwhile'. The rich were thought to be less affected by rationing than 'ordinary people' because they could eat at 'expensive restaurants', 'buy high priced goods ... such as salmon and game' and received 'preferential treatment in shops'. The possibility of surrendering ration coupons in return for restaurant meals was considered but rejected because this policy conflicted with the aim to expand communal feeding. A 5s. price limit on restaurant meals was introduced to quell resentment against so-called luxury feeding. However, this restriction was 'said to arouse "sarcastic comment"', price regulations were easily circumvented and grumbling persisted.[30]

The Ministry of Agriculture called on people to 'dig for victory', but the opportunity to grow vegetables and keep poultry or raise rabbits was not equally available. While allotment keeping was an integral aspect of working-class culture, access to suitable land was much greater among the wealthier income groups in a highly urbanized country such as Britain. Significant food production also required time, which was less available to war workers on long shifts or busy housewives. This is illustrated well by the example of Virginia Potter, an upper middle-class American woman who spent the war living just outside Windsor. With the help of domestic staff, she transformed her garden into a smallholding, devoting over an acre to food production. Potter grew 18 types of vegetable, five varieties of fruit, raised poultry and rabbits and kept a pig with neighbours. In one week in 1942 she produced over 40 eggs at a time when consumption stood at about one per week. Potter frequently entertained guests, bartered, used food as gifts and she sold some of her produce to the local British Restaurant. As a mother of a young child, Potter was exempt from war service and she was able to utilize financial resources to augment her family's rations considerably. However, Potter's 'food production and processing organization' was only possible at the expense of her leisure and her 'life was transformed completely'.[31] Potter's activities were not illegal, but the legal status of barter and gifts of food was ambiguous. Officially rations could be shared within the household, but not among friends or neighbours. MF officials justified this policy in view of the shipping shortage and the 'impossibility ... of proving that money [had been] passed'. The issue became controversial at the

---

29   Rhodes James 1967: 272, 325, Nicolson 1967: 138, 249.

30   TNA, INF 1/292, 16–23 March 1942, 26 May–2 June 1942, 9–16 June 1942, Hammond 1951: 288–93.

31   Brassley and Potter 2006: 226–31, 237, 239.

end of 1942 and, following Winston Churchill's intervention, the regulations were amended to permit 'gifts' of rationed food.[32]

The Black Market

The black market should not be understood merely in terms of theft and the receiving of stolen foods, but rather as endemic circumvention of the emergency legislation.[33] Rationing and food controls, which effectively abolished the price mechanism as the central determinant of economic transactions, confronted many citizens with the prospect of law-breaking for the first time. While there was no large-scale organized black market operated by professional criminals, there is extensive evidence of widespread infringement of the regulations by producers, distributors and retailers, ultimately sustained by public demand.

It was impossible to police thousands of control orders and millions of transactions and the real solution to the black market was to design a rationing scheme that would effectively control itself. A good example is tea, rationed in July 1940. The rationing scheme was based on very loose control of retailers and by 1942 over 8 million rations of tea in excess of the theoretical maximum were being released. This situation was transformed following a recasting of the rationing scheme, which required retailers to surrender coupons to replenish stock. In the wake of this reform, retailers' demand for stock fell below the maximum, saving over 1 million pounds of tea a week.[34]

Some commodities were more difficult to control than others. The black market in meat and practices such as illicit slaughter and under-the-counter sales of eggs and poultry were never entirely eliminated. The shortage of eggs led to an unprecedented trade in supposed hatching eggs by auction, which amounted to a 'legalized black market', and a corresponding racket emerged with regard to alleged breeding poultry.[35] The black market embodied the dark, frequently unacknowledged, underside of 'fair shares'. The authorities did not lose control of a major commodity and the black market arguably functioned as a useful safety valve. However, it also contributed towards a growing division between public and private morality. Most people probably purchased some items in contravention of the control orders, but those on low incomes were least able to afford extra goods at inflated prices.

---

32   TNA, MAF 99/1184, Minute Churchill to Woolton, 21 November 1942; minute, 21 November 1942, teleprint, 23 November 1942, press notice, 23 December 1942.

33   This topic is discussed in Zweiniger-Bargielowska 2000: 151–77, see also Roodhouse 2006.

34   Hammond 1956: 724–34.

35   Ibid.: 94, 65–102, Hammond 1962: 721–6.

How did the Public respond to Food Propaganda, Innovative Foods and New Initiatives?

Wartime food controls undoubtedly transformed the British diet, but it is important to distinguish between policy, propaganda and public response as illustrated by the failure of nutrition education and the low take-up of welfare foods. The most important propaganda tool, the *Kitchen Front* broadcast daily on the BBC just after the morning news bulletin, attracted a sizeable audience. Some new products such as dried eggs were popular at a time when fresh eggs were scarce, but extensive nutrition advice aimed to encourage healthier food choices and cooking methods largely fell on deaf ears. In the face of advice such as steaming vegetables or leaving potatoes in their skins, a survey conducted in 1942 revealed that only one in six followed the latter advice and, as Nicholas put it, British housewives persisted in 'assassinating vegetables'.[36] Similarly, another survey noted that 'large numbers of people have no scientific knowledge of dietetic food values. They consider the foods which make up their traditional diets as those which are good for them'.[37]

This conservatism is illustrated well by the public response to canteens and brown bread. Despite extensive promotion of communal feeding facilities, people generally preferred to eat at home. Only one in five of workers who had access to a canteen actually used the facility, few ate regularly at British Restaurants and people consumed only 2.5 to 3 out of 27 meals per person per week away from home.[38] The nutritional advantages had been advocated by some nutritionists for a long time, but compulsory brown bread was introduced primarily in order to save shipping space and from 1942 the extraction rate stood at 85 per cent.[39] Despite extensive propaganda lauding its health benefits, consumers never really embraced National Wheatmeal Bread. A survey among the general public revealed that three-quarters accepted the loaf 'under wartime conditions', but two-thirds preferred white bread and 63 per cent did not want to see the bread continued after the war.[40] In stark contrast with the popularity of food rationing generally, only half of housewives approved of the bread, which was most unpopular among older and working-class respondents.[41] There is no evidence of a shift in attitudes over time and white bread 'replaced national bread almost completely' when flour was decontrolled in 1955.[42]

Under the welfare foods scheme pregnant and lactating women and young children were entitled to additional milk and certain priority rations. From 1942

---

36   Nicholas 1996: 82–5, Oddy 2003: 153–4, Vernon 2007: 139–46, 223–9.

37   TNA, RG 23/9A, Wartime Social Survey, Food during the War, February 1942–October 1943.

38   TNA, RG 23/9A; MAF 156/396–7. See Chapter 11 by Peter Atkins.

39   Oddy 2003: 138–40.

40   TNA, RG 23/61, Wartime Social Survey, National Wheatmeal Bread, April 1944.

41   TNA, RG 23/9A.

42   Buss 1993: 124–5.

onwards these included subsidized or free vitamin supplements such as orange juice, cod liver oil and vitamin tablets. This scheme was not a great success and take-up of orange juice peaked at just over 50 per cent, about a third took the tablets and cod liver oil was taken by less than one in five of those entitled.[43] In the face of 'continuous publicity and educational efforts' cod liver oil remained unpopular and MF officials recommended that parents should be 'careful to hide any personal dislike' when feeding children the oil.[44] A mother who did not consume all her priority allowances 'lost' the extra calcium and vitamins 'so necessary for maternal and infant health' and according to the official history the take-up was 'disappointing'.[45] This indifference was not distributed evenly. Take-up of vitamin supplements was determined by income group, education level and the age of the mother. Consumption of additional vitamins was significantly higher among the middle classes.[46]

Vital Statistics and Public Health

Despite extensive air raids, there was a considerable improvement in vital statistics among the civilian population during the war. From 1940 to 1951 the standardized female mortality rate declined by 26.3 per cent. Between 1939 and 1945 the maternal mortality rate registered a steep decline from 3.13 to 1.80 per 1,000 births and the infant mortality rate fell from 51 to 46.[47] While diet was only one of several variables which contributed towards these improvements, according to the official history the comprehensive food policy coupled with priority schemes 'did more than any other single factor to promote the health of expectant mothers and young children'.[48] All social groups benefited. A closer look at the statistics in England and Wales reveals that excess mortality among babies from Registrar-General Class V (the poorest), which was more than double the rate of Class I in 1939, remained 'virtually the same' in 1951.[49] A study of rates in Scotland noted that improvement was concentrated among Classes I–III and Classes IV and V showed only 'slight' advances between 1939 and 1945. Indeed, with regard to the post-natal (1–12 months) mortality rate, the social gradient was 'steepest' and the rate for Class V had actually 'slightly deteriorated' during the period. These findings were the 'reverse of expectations' and this evidence draws attention to

---

43   TNA, MAF 75/89, Welfare Foods, appendix 24.

44   TNA, MAF 98/60, Welfare Foods Scheme, Note upon Cod Liver Oil, 29 January 1946; Minute, 14 March 1949.

45   MacNalty 1953: 130–31.

46   Nuffield College Library, Oxford, Wartime Social Survey, Food Supplements: An Inquiry for the Ministry of Food into the Use of Fruit Juices and Cod-liver Oil, 1944.

47   British Medical Association 1950: 93, Winter 1986: 166.

48   MacNalty 1953: 131.

49   Winter 1983: 246.

the limits of class leveling achieved during the war.[50] An internal MF analysis of the relationship between nutrition and infant mortality rates similarly noted that the 'lower social classes should have benefited more than the higher, and the gap should have closed. This, in fact, was not the case'. This disappointing result after years of 'fair shares' and welfare foods was attributed to the fact that the 'higher social classes used vitamin supplements more', a practice which 'may have offset the equalizing effect of rationing'.[51]

The MF took pride in the fact that the 'general health of the civilian was good' throughout the war and the 'fitness of babies and school children was particularly striking'.[52] Vital statistics are a crude measure of nutritional status and public health, but the trends were confirmed by anthropometric data such as children's height and weight. These overwhelmingly registered improvements, but the data also demonstrated the persistence of differentials between children from diverse social backgrounds. According to Sir Arthur MacNalty, the Chief Medical Officer, these were 'greater than can be accepted with equanimity'. Despite these caveats, MacNalty considered the fact that the 'state of the public health should be as good as it is today' despite 'unprecedented strain' as a 'miracle' in 1946.[53] Much of this data focused on children, evidence of morbidity among the adult population was rather more limited and the BMA's *Report of the Nutrition Committee* lamented that 'No significant information is available ... on changes in the physique of adults'. A variety of biochemical studies noted 'uneven' levels of protein, iron and various vitamins which may 'imply uneven bodily reserves, but not necessarily an uneven state of health'.[54]

It was difficult to develop accurate methods to measure nutritional status and surveys of haemoglobin content and body-weight proved to be inconclusive. A Medical Research Council study of heamoglobin content from 1942 onwards concluded that the men examined fell within the 'normal' range, although occupations such as agricultural workers showed levels 'low enough to merit further study'. However, with regard to women and children it was 'impossible to say with any degree of accuracy what constituted the normal level', because standardized instruments had not been available in the past and factors such as menstruation may affect haemoglobin levels. A tentative conclusion merely suggested that there was 'no greater degree of anaemia' than had been found among the 'limited' numbers examined before the war.[55]

The MF conducted body-weight surveys from 1943 onwards with the aim to provide an 'independent check on the adequacy of the national food supply'. Weight trends showed a slight rise in the final years of the war, a decline between

---

50   British Medical Association 1950: 73.
51   TNA, MAF 256/213, Notes on Welfare Foods, n.d.
52   Ministry of Food 1946: 49.
53   Ministry of Health 1946: 1, 120–21, Harris 1995, 165–9.
54   British Medical Association 1950: 81, 89–90.
55   Ministry of Health 1946: 120.

1945 and 47 coinciding with ration cuts and food shortages, and further gains until the study was discontinued in 1949. It was impossible to draw conclusions from these findings because 'no data are available to determine' whether the increases in weight observed were 'normal' for adults. It was also impossible to take into account 'changes in bodily activity' or factors such as loss of sleep and worry in the final years of the war which witnessed renewed air raids and the appearance of flying bombs. Nevertheless, the data provided 'at least an approximate index of the adequacy of the diet' although the weight loss was 'too slight to have had any direct detrimental effect on health'. During the early postwar years, the 'greatest fall in weight' was observed among housewives and workers with no access to canteen facilities and this period was also distinguished by extensive public discontent with the food situation which had negative implications for morale and productivity.[56]

Conclusion

Food policy in Britain during the Second World War was successful. Civilian health was maintained, morale did not break down and Britain won war. People generally considered themselves to be well fed, despite a dramatic reduction in food imports, although the wartime diet was also drab and monotonous. Nevertheless, wartime food policy fell short of the ideal of equality of sacrifice and at times social solidarity was severely strained. The diet of the poorest sections of the working class undoubtedly improved, but wartime food policy did not really establish fair shares. The wealthy and middle classes who had access to land or sizeable gardens, could afford expensive unrationed foods and frequent restaurant meals were least affected by food shortages. At the same time, the poorer sections of the working class continued to be disadvantaged. This was due to limited funds for costly unrationed items, less opportunity to dig for victory and a reluctance to use canteens or take up vitamin supplements. The unequal impact of wartime food policy was relevant not only with regard to class but also gender. Male manual workers suffered most under the flat-rate rationing scheme and it was impossible to ensure that women actually consumed all of their rations. Housewives bore a disproportionate burden in the implementation of wartime food policy and their role as a buffer in the domestic and wider economy extended across the social spectrum. It was difficult to change cooking habits and culinary preferences. There was virtually no evidence of deficiency diseases but nutritional status was never adequately monitored. Mortality rates and anthropometric data registered considerable improvements but the combination of rationing and welfare foods did not eliminate or even reduce class differences in vital statistics.

Rationing and food controls functioned adequately during the war, but the policy was difficult to sustain in peacetime when public willingness to sacrifice

---

56   Harries and Hollingsworth 1953: 75–8, Ministry of Health 1946: 120–21.

was less forthcoming and the purposes of the policy were less clear. In the face of ration cuts housewives' satisfaction evaporated and there is ample evidence of low food morale after the war. As argued elsewhere, the continuation of the austerity policy became politically controversial and extensive discontent, particularly among women, ultimately played a decisive role in the demise of the postwar Labour government.[57]

References

Addison, P. 2005. The impact of the Second World War, in *A Companion to Contemporary Britain, 1939–2000*, edited by P. Addison and H. Jones. Oxford: Blackwell, 3–22.

Brassley, P. and Potter, A. 2006. A view from the top: social elites and food consumption in Britain, 1930s–1940s, in *Food and Conflict in Europe in the Age of the Two World Wars*, edited by F. Trentmann and F. Just. Houndmills, Basingstoke: Palgrave Macmillan, 223–42.

British Medical Association. 1950. *Report of the Committee on Nutrition*. London: British Medical Association.

Burnett, J. 1979. *Plenty and Want. A Social History of Diet in England from 1815 to the Present Day*. 2nd edn. London: Methuen.

Buss, D.H. 1993. The British diet since the end of food rationing, in *Food, Diet and Economic Change Past and Present*, edited by C. Geissler and D.J. Oddy. Leicester: Leicester University Press, 121–32.

Fielding, S., Thompson, P., and Tiratsoo, N. 1995. *'England Arise!': The Labour Party and Popular Politics in 1940s Britain*. Manchester: Manchester University Press.

Hammond, R.J. 1951. *Food* volume I *The Growth of Policy*. London: HMSO.

Hammond, R.J. 1956. *Food* volume II *Studies in Administration and Control*. London: HMSO.

Hammond, R.J. 1962. *Food* volume III *Studies in Administration and Control*. London: HMSO.

Hammond, R.J. 1954. *Food and Agriculture in Britain, 1939–45: Aspects of Wartime Control*. Stanford, CA: Stanford University Press.

Harries, J.M. and Hollingsworth, D.F. 1953. Food supply, body weight, and activity in Great Britain, 1943–9. *British Medical Journal*, 10 January, 75–8.

Harris, B. 1995. *The Health of the Schoolchild: A History of the School Medical Service in England and Wales*. Buckingham: Open University Press.

Mackay, R. 2002. *Half the Battle: Civilian Morale in Britain during the Second World War*. Manchester: Manchester University Press.

---

57   Zweiniger-Bargielowska 2000: 116–17, 124–7, 203–55.

MacNalty, Sir A.S. ed. 1953. *The Civilian Health and Medical Services* vol I *The Ministry of Health Services; Other Civilian Health and Medical Services*. London: HMSO.

Ministry of Food. 1946. *How Britain was Fed in War Time: Food Control 1939–1945*. London: HMSO.

Ministry of Health. 1946. *On the State of Public Health During Six Years of War: Report of the Chief Medical Officer of the Ministry of Health, 1939–45*. London: HMSO.

Nelson, M. 1993. Social class trends in British diet, 1860–1980, in *Food, Diet and Economic Change Past and Present*, edited by C. Geissler and D.J. Oddy. Leicester: Leicester University Press, 101–20.

Nicholas, S. 1996. *The Echo of War: Home Front Propaganda and the Wartime BBC, 1939–45*. Manchester: Manchester University Press.

Nicolson, H. 1967. *Diaries and Letters 1939–1945*. London: Collins.

Oddy, D.J. 2003. *From Plain Fare to Fusion Food: British Diet from the 1890s to the 1990s*. Woodbridge, Suffolk: Boydell Press.

Oren, L. 1973. The welfare of women in laboring families: England, 1860–1950. *Feminist Studies*, 1, 107–25.

Rhodes James, R. ed. 1967. *Chips: The Diaries of Sir Henry Channon*. London: Weidenfeld.

Roodhouse, M. 2006. Popular morality and the black market in Britain, 1939–1955, in *Food and Conflict in Europe in the Age of the Two World Wars*, edited by F. Trentmann and F. Just. Houndmills, Basingstoke: Palgrave Macmillan, 243–65.

Rose, S.O. 2003. *Which People's War: National Identity and Citizenship in Wartime Britain 1939–1945*. Oxford: Oxford University Press.

Ross, E. 1993. *Love and Toil: Motherhood in Outcast London, 1870–1918*. Oxford: Oxford University Press.

Smith, H.L. 1996. *Britain in the Second World War: A Social History*. Manchester: Manchester University Press.

Seymour, J. 1992. 'No time to call my own': women's time as a household resource. *Women's Studies International Forum*, 15(2), 187–92.

Spring Rice, M. 1939. *Working-Class Wives: Their Health and Conditions*. Harmondsworth: Penguin.

Twigg, J. 1983. Vegetarianism and the meanings of meat, in *The Sociology of Food and Eating: Essays on the Sociological Significance of Food*, edited by A. Murcott. Aldershot: Gower, 18–30.

Vernon, J. 2007. *Hunger: A Modern History*. Cambridge, MA: Harvard University Press.

Winter, J. 1983. Unemployment, nutrition and infant mortality in Britain, 1920–50, in *The Working Class in Modern British History: Essays in Honour of Henry Pelling*, edited by J. Winter. Cambridge: Cambridge University Press, 232–55.

Winter, J. 1986. The demographic consequences of the war, in *War and Social Change: British Society and the Second World War*, edited by H.L. Smith. Manchester: Manchester University Press.

Zweiniger-Bargielowska, I. 2000. *Austerity in Britain: Rationing, Controls, and Consumption, 1939–1955.* Oxford: Oxford University Press.

Chapter 11

Communal Feeding in War Time: British Restaurants, 1940–1947

Peter J. Atkins

Introduction

When asked to comment on the London County Council's (LCC) plans for a history of wartime efforts to feed the capital's blitzed population, one insider commented that 'the story is worth telling .. we are recording an epic in history'.[1] Although this history was unfortunately never published, for subsequent generations food has always played an important part in imagining the experience of the nation at war.[2] Much of the literature has focused upon the supply chain ('dig for victory', 'the national farm', import shortages) or rationing and its impact upon diet and nutrition. This leaves a gap for the present paper in the area of communal feeding. I will look at the curious and somewhat misunderstood institution of the British Restaurant (BR), which operated from 1941 to 1947 and arguably achieved notoriety far beyond its numerical significance. In 1942 one commentator perceptively observed that BRs 'may be said to have started as an improvisation and to continue as a compromise'.[3] The implication of this statement is of a lack of strategic foresight, yet there were some positive outcomes that are worth looking at, and also some unintended consequences.

This chapter is divided into four parts. First, the origins and development of BRs are analysed, particularly with regard to the rhetoric and hidden purposes of the Ministry of Food (MF) and of political interests generally. Second, I will briefly introduce a regional perspective, which, as far as I am aware, has not been attempted before. Third, I will show that pulling together for the war effort was not a feature of the catering sector, where vitriolic criticism was made of the government's communal feeding policies. Fourth, there is consideration of the food served in BRs.

The historiography of BRs is interesting in its own right. R.J. Hammond in his official three volume history of wartime food control devotes a whole chapter to

---

1   London Metropolitan Archives (LMA), LCC/RC/GEN/1/1: E.A. Hartill, 13 January 1944.

2   Although the LCC history remained in draft, the official war history did provide three volumes on food, authored by R.J. Hammond.

3   Anon. 1942: 675.

the restaurants and this remains the most detailed account. Hammond's approach shows a welcome irreverence towards the decision-making process of government and reveals tensions and rivalries within and between ministries. There is a degree of what one might call 'creative chaos under fire' in his narrative, especially in the early years when air raids threatened to cause widespread dislocation. Writing in the 1950s, Hammond presumably had access to the relevant civil servants and their 'inside stories', and certainly some of his interpretations go well beyond the evidence that has survived in the papers of the MF. Since Hammond there has been little of a critical nature written about BRs, although we have a contextualized commentary by Ina Zweiniger-Bargielowska and, more recently, a book by James Vernon that touches on communal feeding.[4]

Origins and Development

According to the official war history, early plans for emergency feeding were inchoate. The idea of reviving the National Kitchens that had figured in the First World War was soon dropped. There appears to have been some bickering between ministries about who should take on the responsibility of feeding in the event of enemy attacks. In the spring of 1940 the advent of Lord Woolton as Minister of Food, and then Churchill as Prime Minister, was something of a turning point. By July an experiment was being conducted by the MF on a working-class housing estate in North Kensington.[5] Over 2,000 hot meals per week were cooked on simple ranges, the choice being limited to popular dishes such as Irish stew and dumplings, or roast beef. A different main meal was cooked each day at an affordable price by volunteer labour. People's reactions seem to have been largely positive, although timeliness was identified as a key issue because workers and school children all needed to eat quickly in the short lunch period available to them. The ministry was sufficiently encouraged by this project to envisage the scaling up of catering to meet local needs in what were to be called Community Feeding Centres.

In early September Woolton requested that the LCC should take the lead in providing communal feeding facilities in the capital.[6] The departure point was the need to help people unable to prepare meals for themselves due to temporary interruptions of gas, water and electricity services because of bombing.[7] These emergency facilities were important, in the words of Richard Titmuss, for

---

4   Zweiniger-Bargielowska 2000, Vernon 2007.

5   Gates 1942: 102, The National Archives, Kew (TNA), MAF 99/1797, 'Ministry of Food, the Communal Restaurant: an Experiment', October 1940.

6   The first approach was on 10 September, just after the first major air raids, with the formal letter following a week later. The Minister guaranteed that the Council would not be out of pocket as a result of this policy, LMA, LCC/RC/GEN/1/1.

7   London County Council, Civil Defence and General Purposes Committee, 'The Londoners' Meals Service', 21 October 1940.

absorbing the shock of air raids.[8] Called the Londoners' Meals Service (LMS), this was always separate from the BRs scheme. But in effect the two were similar, at least in post-blitz London, which was dominated by field kitchens and mobile canteens offering a 'cash and carry' service. The first indoor dining room was opened in Woolmore Street, Poplar on 24 October 1940.[9] By Christmas, 139 LMS centres were producing a total of 80,000 meals a week. Many of the sites were schools, first because the buildings were increasingly available as children were evacuated, and, second, because the domestic science teachers and their facilities would otherwise have been unemployed and underutilized. The pricing formula was 'cost of food + 25 per cent + ½d for fuel', working out at an affordable 9d or 10d for a two-course meal.

In November 1940 provincial local authorities were circulated, asking them to consider setting up what were now to be called Community Kitchens.[10] By the end of the year these had been established in major cities such as Birmingham, Bradford, Leeds, Liverpool, Manchester and Newcastle upon Tyne. Progress was slow at first but the spread of air raids concentrated the minds of councils, as did the minister's offer of financial assistance.[11]

Churchill disliked terminology such as 'Communal Feeding Centre' and 'Community Kitchen' as redolent 'of Communism and the workhouse'. In March 1941 he suggested the name 'British Restaurant' because the word 'restaurant' is associated positively in people's minds with 'a good meal'.[12] One modern branding professional sees this in retrospect as the masterstroke of someone who instinctively understood the difference between product and brand.[13]

The process of setting up BRs was fairly bureaucratic.[14] At first the ministry insisted on approving all applications from the centre and the paperwork often took months, involving the allocation of equipment,[15] requisition of buildings and recruitment of staff.[16] To short-cut this process, some local authorities decided to open their own communal restaurants, as did voluntary organizations such as

---

8   Titmuss 1950: 346.

9   LMA, LCC/RC/GEN/1/1, 'LCC, Meals Services, Origin of the Service, [1944]'.

10   TNA, MAF 83/382, MAF 99/1796.

11   TNA, MAF 74/49, Ministry of Food, Public Relations Division, Information Branch, 'British Restaurants', 3 September 1943.

12   Memo to Minister of Food, 21 March 1941, Churchill 1950: 663.

13   Bernstein 2003: 1137–8.

14   Some sample documents have been preserved for Barrow-in-Furness and other places, see TNA, MAF 99/1684–6.

15   Equipment was scheduled under 150 different headings, including solid fuel ranges, as well as electric and gas cookers; potato peeling machines; electric washing machines; refrigerators and insulated containers; sinks, scales, saucepans and furniture, TNA, MAF 74/49.

16   From May, 1941, Divisional Food Officers were given this power.

the National Council of Social Service and the Women's Voluntary Services.[17] The advantage of being inside the official system was that all capital costs were reimbursed. The disadvantage was that ministry officials continued to micro-manage, such as suggesting menus, monitoring food quality and insisting on each outlet being financially self-supporting.[18]

Most BRs were run on the cafeteria principle.[19] The diners bought tickets and then queued up and chose food from a series of hot plates. From May 1941 onwards a number of cooking depots were set up around the country in order to supply food in bulk to the BRs and schools in that locality. My estimate is that about 10 per cent of BR meals were supplied in this way and, surprisingly perhaps, the quality was said to have been indistinguishable from the meals prepared on site.[20]

BRs received allowances for rationed foods on the same scale as commercial catering establishments, although the quantities were higher where at least 60 per cent of the clientele were industrial workers, especially for those in Category B – heavy manual labour (Table 11.1).[21] BRs were just one element of a broad government wartime food policy, which can be divided into the systematic (rationing, welfare foods, milk in schools) and the practical. The latter included provisions for day-to-day feeding (BRs, school canteens, factory and pithead canteens, and a rural pie scheme) and emergency feeding (cooking depots, emergency meals centres, rest centres, air raid shelter canteens, Queen's Messenger Convoys, and other mobile canteens).[22] This system was administered by three ministries, namely Food, Education and Labour.

As a result of this complexity, the term 'British Restaurant' was confusingly vague. We have already mentioned the dining rooms set up under the MF's scheme. These were supplemented by the LCC's LMS, by other local authority schemes and by restaurants set up by voluntary organizations. All of these counted in official statistics as BRs but they often had no direct connexion with the government. In addition, evacuee feeding centres were sometimes rebranded as BRs, as were school canteens that served meals on a daily basis to the general public.[23]

---

17   For instance Bournemouth, Eastbourne, Hastings, Hull, Newcastle, Oxford and Wolverhampton.

18   Food quality was a sensitive issue. Woolton was anxious for his staff to remember 'the Ministry's prestige was very closely associated with the efficiency of British Restaurants [and] he was anxious that the quality of the service and other meals served should be maintained at a higher level', TNA, MAF 99/1716, memo by Mr Harwood, 27 October 1941.

19   TNA, MAF 74/49.

20   TNA, MAF 99/1734, City of Birmingham, Reconstruction Committee, 'British Restaurant Enquiry, September 11 to October 6, 1944'.

21   Pyke 1944b: 231, TNA, MAF 74/49.

22   Jones 1944: 121–40, Ministry of Food 1946: 43–5.

23   These were the result of deals done with the Ministry of Health and the Board of Education, TNA, MAF 74/49.

Table 11.1   Catering Allowances per Main Meal, 1943

Food	Normal caterers	Allowances for industrial workers	
		Category A	Category B
Bacon (oz.)	0.14	0.14	0.14
Fats (oz.)	0.30	0.50	0.50
Sugar (oz.)	0.12	0.12	0.20
Meat (oz.)	1.00	1.50	2.00
Fish (oz.)	0.32	0.32	0.32
Cheese (oz.)	0.21	0.21	0.21
Preserves (oz.)	0.14	0.14	0.14
Dried egg (oz.)	0.16	0.16	0.16
Liquid milk (pts)	–	–	–
Skim milk powder	0.12	0.12	0.12
Sausage meat (oz.)	0.67	0.67	0.67
Suet (oz.)	–	0.08	0.08

Source: TNA, MAF 256/197.

Rhetoric and Purpose

Why did the government favour BRs? Apparently, the MF had envisaged 10,000 restaurants spread around the country, but it achieved only 1,500 under their own scheme, and a peak of 2,160 overall.[24] Essentially public rhetoric and private memoranda employed two types of argument. First, there was a cluster that we might call functionalist or utilitarian justifications. There were suggestions, for instance, that BRs served the war effort by improving efficiency in one way or another. In 1941 their purpose was said to be principally:

> To ensure that people who, owing to war conditions, have difficulty in securing meals, shall be able to have a least one hot nutritious meal a day at a reasonable

---

24   The Ministry began pressurizing local authorities in 1940 but this ceased in 1943, TNA, MAF 99/1759.

price. Such people include those whose incomes have fallen, old age pensioners, and others with small fixed incomes, women engaged in war work, men whose wives and families have evacuated, and evacuated persons who have difficulties owing to limited domestic accommodation. School children are also catered for in a number of restaurants and this service is likely to expand very considerably.[25]

A possible reading of this statement is that BRs were a form of infilling where factory canteens were not provided, for instance in industrial districts dominated by workshops, and where local education authorities were not supplying school dinners.

Related to this was an economy of scale argument. Resources of various kinds were of course in short supply in wartime and BRs were said to economise on fuel to cook meals and labour to prepare and serve them.[26] Hidden beneath was the point that, where communal facilities were available within easy walking distance, it became difficult for housewives to resist the call to work on the grounds that their domestic labour was irreplaceable. Nutrition was also frequently cited as a justification for government-sanctioned feeding schemes. Dieticians were used in formulating menus and the ministry deployed scientific expertise to analyse the content of meals. BRs were therefore a small cog in the larger engine of food policy that strove to improve health and working efficiency.

Second, the political case for BRs was partly ideological and partly tied to wartime strategy. The first element was the subject of an unseen struggle in the wartime coalition government between Conservatives, such as Woolton, and those on the left. The latter constantly stressed that 'the restaurants are used mainly by the working classes and the lower paid professional and clerical classes'.[27] Ernest Bevin, the Minister of Labour, frequently demanded an expansion of industrial canteens,[28] whereas Woolton maintained that 'there is ... no restriction in admittance to British restaurants; they are open to all members of the public'.[29] Woolton's justification for this was that BRs were intended for those involved in war work, and that this was not restricted to fighting or making munitions. They should therefore equally be open to shop assistants, office workers and housewives. Others, from the right, saw communal feeding as 'entirely abhorrent to the British

25   TNA, MAF 99/1589, Ministry of Food, 'Memorandum on British Restaurants', [1941].

26   TNA, MAF 74/49, Ministry of Food, Public Relations Division, Information Branch, 'British Restaurants', 3 September 1943.

27   TNA, MAF 99/1590–94, monthly reports on British Restaurants to the War Cabinet.

28   Hammond 1956: 390.

29   TNA, MAF 99/1589, Ministry of Food, 'Memorandum on British Restaurants', [1941].

way of life' and this divide was later to be a live political issue when the war ended but BRs continued.[30]

A decision was made early on in the war not to close down commercial restaurants or to charge the food they served against people's rations.[31] Following on from this there was the oft heard accusation of waste and 'luxury feeding' in expensive restaurants. In a sense, BRs were a balancing measure, giving equivalent access, off the ration, to people who would otherwise have been unable to afford to eat out. There was a deliberate policy to make eating in a BR an uplifting experience. The décor was lightened and even details such as the font of the lettering on notices were discussed. A few restaurants had live music and many had art, either newly painted murals or specially chosen prints. In short, here was a vehicle for raising morale. BR customers, according to a survey in Birmingham, seem to have appreciated the food, the service, and the 'homely' atmosphere.[32]

London and the Regions

In 1942 most local authorities with populations over 50,000 (mostly County and Municipal Boroughs) had adopted the BR idea. In the band 10,000 to 50,000 it was about a half, and a quarter for those authorities under 10,000.[33] Twelve local authorities had ten or more restaurants open each, and London dominated with a quarter to a third of BRs nationally.[34] Table 11.2 shows regional variations at the scale of the Food Office District. For the sake of comparison, some data is included on commercial catering premises from a census by the MF in 1940.

Although it was anticipated that the enemy would bomb vital installations and maybe civilian targets, plans to deal with the consequences were slow to recognize the need to feed displaced populations. Communal feeding in various guises was encouraged but the MF throughout the war avoided centralized compulsion. Instead they relied upon persuading local authorities to take responsibility for the particular circumstances of their area. This amounted to a redefinition of the role of the local state.

30   Ernest Burdett in Morgan et al. 1946: 515.

31   Woolton 1959: 220. This was different from the decision made in Germany to deduct café meals from ration quotas, Anderson 1943: 27.

32   TNA, MAF 99/1734, A 1944 survey of the British Restaurants in Birmingham found that the vast majority of the 1530 people questioned were favourably disposed.

33   TNA, MAF 152/55.

34   TNA, MAF 74/49.

Table 11.2   The Regional Pattern of British Restaurants and Civic Restaurants

	Population per catering establishment, 1940	Population per British Restaurant, 1941	Numbers of British Restaurants		Civic Restaurants, 1948
			1941	1943	
England					
Eastern I	308	14,659	37	56	22
Eastern II	228	16,061	45	83	25
London	237	13,736	364	501	212
Midlands	251	79,329	34	138	127
North Midlands	308	16,927	67	131	42
North	493	16,432	74	185	31
North East	334	23,335	105	176	64
North West	275	57,832	74	166	77
South	238	19,708	71	197	48
South East	185	8,794	127	112	42
South West	229	16,429	67	108	38
Wales					
North	171	160,000	1	8	4
South	336	22,932	33	84	15
Scotland					
East	331	36,000	5	16	2
North	278	–	0	1	1
North East	407	–	0	6	0
South East	380	30,000	15	21	6
West	537	231,425	8	46	17
N. Ireland	427	34,615	13	14	0
UK	280	20,910	1140	2043	773

Sources: TNA, MAF 74/49; MAF 83/382; MAF 99/519; MAF 99/1589.

Government had previously been concerned with such matters as education, public assistance, parks and playgrounds, and utilities. But more vital matters, such as air raid precautions, life in shelters, the evacuation of children, or the operation of British Restaurants, required a greater understanding of the way of life of the people, their habits and desires, their hopes and fears. The relationship of the local official to the community as a whole in the pre-war period was an important one, but it was generally related to matters that were impersonal to most of the community. The emphasis has now changed to matters of vital personal concern for citizens. [35]

Initially there was some irritation in the Ministry at the attitude of some local authorities. Despite the inducements offered in the form of capital grants, guarantees against operating losses and professional advice on practical details, 'the vast majority' of councils by early 1941 had not welcomed the idea.[36] The reaction was said to have 'varied from true passive resistance to lukewarm acquiescence... The general retort to any approach ... has been that a demand ... does not exist in that particular town.'[37] Town clerks apparently 'seized on any pretext for delay' and were especially exercised by the lack of a clear legal framework for action, for instance in the requisitioning of premises. This excuse disappeared on 28 January 1941 with the making of the Local Authorities (Community Kitchens) Order under Regulation 54B of the Defence (General) Regulations, 1939. Nevertheless the government's non-aggressive policy was restated in a circular letter the following month:

> The Minister of Food does not wish to cause local authorities to set up Community Kitchens when the need does not exist, the intention of the Order is solely to give adequate authority for the establishment of Community Kitchens where there is need for them.[38]

In view of the resistance and apathy in some areas, it is not at all surprising that there was a great deal of geographical variation in implementation. Local politics in Manchester, for instance, were said never to have been favourable to BR.[39] When pressed the city authorities preferred to open outlets in the suburbs rather than in the city centre.[40] By contrast, Liverpool, Birmingham and Bristol bought into the concept at an early date and made substantial local provision. Even London had

---

35   Biddle 1942: 83.

36   French 1943, Q.2855, made it clear that the Ministry considered the route of direct control: 'after all, we are a very large trading organization'.

37   TNA, MAF 99/1589, Ministry of Food, 'Memorandum on the Position of Community Feeding', January 1941.

38   TNA, MAF 99/1609.

39   *Daily Telegraph*, 29 December 1944.

40   TNA, MAF 99/1759, Memorandum, 'Establishment of British Restaurants', [1942].

great diversity. The boroughs varied in their initiative and enthusiasm to the extent that Chelsea and Poplar had one restaurant per 8,000 people, whereas Stepney had one per 70,000.[41] Ellen Leopold attributes this at least partly to civil defence planning which encouraged the oversupply of facilities in west London for the benefit of evacuees who would have gone there from the south coast in the event of an invasion.[42] Towns responded differently to the BR idea. Many were happy with a pared down version that meant trestle tables and benches, while for others the presentation of what they saw as a 'social service' was at the core of their civic pride. An example is the degree to which the price of meals (mostly lunch) were subsidized. In 1944, for instance, the vast majority were charged at 8d or 9d, but a quarter of authorities opted for less and some insisted on as much as 1s.[43]

Commercial Resistance to Civic Entrepreneurship

One explanation for geographical variations was the power of chambers of commerce in many localities. On behalf of the catering trade, the chambers opposed central interference in the free market under the cover of war measures. Private caterers objected that they could not produce a meal equivalent to that in BRs at a comparable price. A confidential estimate by the MF in 1946 was that a standard cafeteria meal costing 1s 3d in a BR was at least 1s 10d in a Lyons outlet.[44]

The official war history reveals the advantages enjoyed by BRs.[45] They benefited in effect from interest-free loans and the guaranteed write-off of any operating losses that were not too excessive. Their equipment was purchased centrally. They received professional advice on sites, equipment and food standards from ministry officials. To some extent this was balanced by the fact that many were in unsuitable premises, serving restricted menus, and with costs inflated by the payment of wages approved by the Joint Industrial Council that were above the catering industry norm. Direct comparisons with the private sector are therefore difficult. On 12 January 1942 Woolton met with a deputation from the catering trade. He promised to look at representations about proposals for any new restaurants that were said to be unnecessary in view of existing commercial provision. This was repeated in an answer to a parliamentary question two weeks later.[46]

Profitability was variable. In the financial year 1942–3, after allowing for the amortization of capital, 698 local authorities running BRs achieved a net profit and this was repeated in 1943–4.[47] After the war, Gilbert Sugden found that

---

41   LMA, LCC/RC/GEN/1/26.
42   Leopold 1989: 208.
43   TNA, MAF 99/1797.
44   TNA, MAF 99/137, 'Brief for the Minister', 21 May 1946.
45   Hammond 1956: 397–8.
46   *Hansard* 377, 28 January 1942, c. 717.
47   TNA, MAF 99/1609.

civic restaurants were still mostly profitable in 1947–8, although some care is needed with his conclusions because authorities running loss-making portfolios of restaurants were forced to close them down.[48] This happened most famously to the LCC, whose costs soared, particularly due to rising rents in the city centre.

Opinions about alternatives were explored in the wartime social survey. In February 1943 a stratified sample of 4490 industrial workers found that 42 per cent had lunch at home, 22 used a canteen, 19 per cent ate sandwiches, and 11 per cent frequented cafés.[49] A 1944 survey of BR customers in Birmingham found that 62.6 per cent of respondents saw going home as their main option, and 11.2 per cent would have eaten sandwiches. Only 3.8 per cent considered a private restaurant or café.[50] Convenience seems to have been a major factor since over half of customers travelled five minutes or less for their meal and 91 per cent for 15 minutes or less. Clearly this would not have been possible in cities with fewer outlets than Birmingham.[51] One argument in favour of BRs was that they had played their part in the enormous increase during the war of eating out. On balance it was therefore likely that they had helped to increase trade for catering generally rather than competing with the private sector.[52]

The Food in British Restaurants

The MF from the outset thought carefully about the nutritional standard of meals served at BRs. In March, 1941, ministry dieticians prepared sets of menus, taking into account regional preferences, such as in Scotland.[53] The same year a booklet entitled *Canteen Catering* was issued. It listed standard and special recipes, with suggestions for alternatives where supplies were short or variable.

Generally speaking, the food in BRs was said to be of good quality and filling.[54] There were some attempts to introduce meals in the Oslo style, with the intention of providing in one sitting all of the day's needs for animal protein, vitamins and minerals.[55] But this met with resistance from customers who wanted their

---

48   Sugden 1949.

49   Box and Thomas 1944: 162.

50   These data are at odds with a London survey in 1943, where the percentages were 24, 27, and 18 respectively. No doubt the longer commuting distances in the big city will have been a factor. London Council of Social Service 1943: 17.

51   TNA, MAF 99/1734, City of Birmingham, Reconstruction Committee, 'British Restaurant Enquiry, September 11 to October 6, 1944'.

52   TNA, MAF 99/1734, National Council of Social Service, Report of the Conference, 'The Future of Communal Restaurants', 7 February 1944.

53   TNA, MAF 74/49.

54   They were said to be superior to those served in the restaurants of the Sorbonne, in Paris, *The Times*, 22 February 1947, 6.

55   Pyke 1944a: 92.

traditional meat and two vegetables.[56] In Birmingham all 56 BRs had a choice of five meat dishes, five vegetables and five desserts, and those in the city centre had more. In other cities with less on offer, menus had to be removed from the entrances because customers would 'wander from one to another and the restaurant serving roast attracted the customer'.[57]

A meeting was held in June 1942 to request the collaboration of universities and research institutes around the country.[58] In the chair, Dr Magnus Pyke, of the MF's Scientific Adviser's Division, suggested a start with work on the vitamin C content of canteen meals. This was because restrictions on fruit intake transferred the onus of delivering vitamin C on to vegetables, and especially cabbage. There was concern that mass catering, particularly the use of hot cupboards, was destructive of this vitamin, so the research results were eagerly anticipated. It had initially been planned that a main meal in a BR would provide one third of the day's energy needs.[59] In practice, the survey found (Table 11.3) about 22 per cent of recommended calories in an average BR lunch. This was partly because the use of potatoes as a substitute for bread gave meals a bulky and unappetizing appearance. Vitamin C was low in winter.[60]

Table 11.3   The Nutritional Content of British Restaurant Meals in February 1943

	Standard	Actual
Energy	1000 k cals	626 k cals
Protein	24 g.	22 g.
Calcium	270 mg.	186 mg.
Iron	8 mg.	4–9 mg.
Vitamin A	200 i.u.	1000 i.u.
Vitamin B$_1$	200 i.u.	136 i.u.
Vitamin C	50 mg.	28–49 mg. (seasonal)
Riboflavin	0.9 mg.	0.3–0.9 mg.
Nicotinic acid	12 mg.	7 mg

Source: TNA, MAF 256/197.

56   *The Times*, 2 July 1942, 2.
57   LMA, LCC/RC/GEN/1/1, memo by M.C. Broatch, 3 September 1943.
58   TNA, MAF 83/382, MAF 98/61.
59   TNA, MAF 256/197, 'The Nutritive Value of Communal Meals', 15 February 1943.
60   Booth et al. 1942.

Conclusion

An occasional trope in the confidential papers of the MF was of sympathy for the plight of women and the promotion of BRs and other forms of communal feeding as means of easing the burden of domesticity. But feminist research, while acknowledging women's vital role in wartime industry, rejects institutions such as communal feeding as of any long-term significance. Their facilitating role was minimal since the expectation upon working women was now of a double burden that included a return to all of the pre-war commitment to cooking and child care.[61]

What then of the other achievements of the MF's BR policy? The functional arguments that I referred to above were modest in their outcome. The best we can say is that at the height of the war about half a million people a day (including children) received a cheap but nutritious meal that supplemented their rations. This filled a small niche in industrial feeding, particularly in the workshop cities such as Birmingham, but maybe less so in factory cities such as Manchester, where works canteens bore the burden.

Three methods of quantifying this impact were used at the time. The first, as used by the MF, was to look at the allocation of rationed foodstuffs such as meat, as a surrogate measure.[62] On this basis it was calculated that, in August 1941, BRs received 3.7 per cent of the catering total. Second, various estimates were made of the number of meals served. Again in 1941, BRs were calculated to have managed only 0.9 per cent of total, with commercial restaurants at 38.3 per cent, and industrial canteens at 14.5 per cent.[63] A 1942 version of the latter, given in a parliamentary answer, revealed somewhat different figures at 1.8, 57.1, and 41.0 per cent respectively.[64] And a retrospective enquiry at the end of the war found that BRs were providing 3.5 per cent of main meals in January 1942, rising to 7.5 per cent by March 1944.[65] The third approach was to ask the consumers where they ate. The wartime social survey in February 1943 found that only 2 per cent ate in BRs.[66] This is probably the most reliable figure. The instability in the data above is due to the definition of a 'meal', which on some occasions included tea or snacks, but on others was restricted to cooked main meals. Overall, we can say with confidence that BRs contributed only marginally to wartime feeding.

The more intangible political considerations are a little more positive. Most importantly perhaps, BRs contributed to a debate about communal feeding that continued after the war, but which ultimately ran into the sand at the mid-1950s

---

61   Summerfield 1983, Jackson 1992: 160.

62   Each main meal in the catering sector was allocated 1d worth of meat and, in calculating total consumption, the ration for domestic consumption could be added.

63   TNA, MAF 83/382, Committee on Catering Establishments.

64   *Hansard* 383, 22 October 1942, c. 2121.

65   TNA, MAF 99/1734.

66   Box and Thomas 1944: 162.

political hinge point with the abolition of rationing in 1954 and entrenchment of Conservative ideals at the 1955 general election.

References

Anderson, C.A. 1943. Food rationing and morale. *American Sociological Review*, 8(1), 23–33.

Anon. 1942. The function of British Restaurants. *Nature*, 149, 675–8.

Bernstein, D. 2003. Corporate branding – back to basics. *European Journal of Marketing*, 37, 1133–41.

Biddle, E.H. 1942. British civilian agencies in the war. *National Municipal Review*, 31(2), 79–83.

Booth, R.G., James, G.V., Marrack, J.R., Payne, W.W. and Wokes, F. 1942. Ascorbic acid in meals at British Restaurants and school canteens. *Lancet*, ii, 569–71.

Box, K. and Thomas, G. 1944. The Wartime Social Survey. *Journal of the Royal Statistical Society*, 107, 151–89.

Churchill, W.S. 1950. *The Second World War. Vol. III: the Grand Alliance*. London: Cassell.

French, Sir H. 1943. Evidence, *Reports from the Committee of Public Accounts*, P.P. 1942–43 (116) II. 285.

Gates, M. 1942. British Restaurants in the North-Western Division, *Public Administration*, 20(3), 101–8.

Hammond, R.J. 1956. *Food. Volume II: Studies in Administration and Control*. London: HMSO.

Jackson, S. 1992. Towards a historical sociology of housework: a materialist feminist analysis, *Women's Studies International Forum* 15, 153–72.

Jones, T.G. 1944. *The Unbroken Front: Ministry of Food, 1916–1944*. London: Everybody's Books.

Leopold, E. 1989. LCC restaurants and the decline of municipal enterprise, in *Politics and the People of London: the London County Council 1889–1965*, edited by A. Saint. London: Hambledon Press, 200–213.

London Council of Social Service. 1943. *The Communal Restaurant: A Study of the Place of Civic Restaurants in the Life of the Community*. London: LCSS.

Ministry of Food. 1946. *How Britain was Fed in War Time: Food Control 1939–1945*. London: HMSO.

Morgan, J., Burdett, A.E., and Hodge, H. 1946. Eating out. *The Listener*, 18 April, 511, 515–16.

Pyke, M. 1944a. Food supplies for collective feeding. *Proceedings of the Nutrition Society*, 1, 1–2, 92–8.

Pyke, M. 1944b. Some principles of war-time food policy. *British Medical Bulletin*, 2, 10–11, 228–31.

Sugden, G. 1949. The finance of civic restaurants: an investigation. *Local Government Finance*, 53, 237–45.

Summerfield, P. 1983. Women, work and welfare: a study of child care and chopping in Britain in the Second World War. *Journal of Social History*, 17, 249–69.

Titmuss, R. 1950. *Problems of Social Policy*. London: HMSO.

Vernon, J. 2007. *Hunger: A Modern History*. Cambridge, MA: Harvard University Press.

Woolton, Lord. 1959. *The Memoirs of the Rt Hon. The Earl of Woolton*. London: Cassell.

Zweiniger-Bargielowska, I. 2000. *Austerity in Britain: Rationing, Controls, and Consumption, 1939–1955*. Oxford: Oxford University Press.

Chapter 12

Rationing and Politics: The French Academy of Medicine and Food Shortages during the German Occupation and the Vichy Regime

Isabelle von Bueltzingsloewen

Introduction

Few historians have studied the ways in which health conditions in France were affected during the Second World War and the German occupation.[1] It is true that it is difficult to evaluate these conditions. France was not subject to the widespread famine that struck Greece or the Netherlands, not to speak of Poland and the Soviet Union. Food shortages, combined with a lack of drugs, shortages of heating fuel and a general decline in sanitary conditions nonetheless led to many deaths among the most vulnerable sectors of the population.[2] A massive number of infants died between 1940 and 1945[3] and the elderly, both those living alone and residents in hospices[4] equally paid a heavy toll due to undernourishment. Other groups were prisoners, inmates of internment camps, mentally ill patients in psychiatric hospitals,[5] the chronically ill and, finally, children who were wards of the state. It is impossible to establish the exact number of 'sanitary victims' (which varied greatly from one department to another),[6] because they were not counted in demographic statistics at the end of the war.[7] We do know, however, that the average annual mortality rate for the period from 1940–45 was 17.4 per

---

1 This lack of interest cannot be exclusively justified by the disappearance of the archives of the State Department for Health and the Family (Secretariat d'Etat à la Famille et à la Santé).

2 von Bueltzingsloewen 2005.

3 Rollet and De Luca 2005.

4 François Chapireau estimates that 50,000 elderly persons died of hunger in hospices, Chapireau 2009.

5 Approximately 45,000 victims have been counted between 1940 and 1945, von Bueltzingsloewen 2007.

6 Certain departments, particularly those in western France, were relatively spared, while others had dramatic food shortages. Significant differences further existed between urban and rural areas.

7 Immediately after the war, demographers estimated the number at about 300–350,000, Bunle 1947, Vincent 1946.

cent, much higher than the rate of 12.3 per cent in 1938. These figures do not fully account for the abnormally high number of deaths caused by deteriorating sanitary conditions, because the increase in the number of deaths due to undernourishment was compensated by a decrease in the number of deaths related to other conditions such as alcoholism, suicide or an overly rich diet.[8]

In contrast with other belligerents which were subject to the blockade, most of France with the exception of occupied departments in the North and East escaped food supply problems during the First World War. However, the food question became critical in France in 1940, or even in 1939 due to the enormous need of supplies as a result of mobilization.[9] In a few short weeks, shortages of all kinds radically transformed everyday life in France. Numerous notebooks and personal diaries published after the war attest to this situation and finding food became a daily source of anxiety, especially in urban areas. A great deal of time and ingenuity was required to find even insipid, poorly-nourishing produce such as rutabagas or cabbages, as well as basic necessities such as cloth, soap, washing powder, coal, petrol, string or bicycle parts.[10]

We have little knowledge of the reaction of the medical profession when confronted with a situation for which it was unprepared: symptoms of undernourishment, a spectacular deterioration in sanitary conditions and an unprecedented decline in medical supplies and treatment options. This chapter examines the response of a section of the medical elite who were members of the Academy of Medicine.[11] In order to remain faithful to their role as 'guardians of public health', the representatives of this prestigious institution established in 1820 found it necessary to distance themselves from the regime of Marshal Pétain, despite the fact that many approved of the authoritative nature of the regime.[12]

Efforts to Alleviate Rationing Policy: An Illusory Strategy

A careful reading of the minutes of the public sessions of the academy from September 1939 to May 1940 indicates that the issue of the predictable consequences of food shortages on public health did not mobilize this body during the so-called

---

8   Death certificates often did not specify the cause of death, Chevallier and Moine 1945.

9   After the German invasion, food supplies deteriorated despite France's agricultural wealth due to disruptions in transport and production along with labour shortages.

10   Veillon 1995, Alary 2006.

11   Other members were biologists, chemists and veterinarians, see website of the Academy of Medicine.

12   It is difficult to consult the archives of the Academy of Medicine because there is no existing inventory. I wish to thank Jérôme Van Wijland for his precious help.

'phoney war'.[13] On 10 October 1939, academy members declared that they were 'at the Government's disposal for any contribution they could make to the country's defence', but they were not consulted during preparations of the March 1940 decree which established rationing in France.[14] The war was nonetheless high on the academy's agenda, which included topics such as improving the care of the wounded, pregnant women and refugee children. Formal talks focused on war wounds, gas gangrene and blood transfusions. However, the prevention of epidemics, the principal mission of the academy since its founding, remained the main preoccupation. Members of the academy pointed out the need for a special diphtheria prevention programme in the case of an epidemic in October 1939 and, again, in March 1940. During this period, they further recommended vaccination of the civilian population against smallpox and typhoid fever and they discussed the need for vaccinating against typhus.

The issue of food rationing was not brought up until the German invasion. In an appeal adopted on 28 May 1940, academy members reported that a sugar ration of 25 g per day was 'altogether insufficient for infants and children under the age of three'.[15] The academy requested that the sugar ration be raised to 50 g and they rejected substitution by saccharin, which had no nutritional value. However, upon a request by the Ministry of Health, the academy authorized the use of saccharin and, subsequently, numerous other substitutes were accepted. In September, a serious epidemic of gastroenteritis caused by bad quality milk given to infants who were not breastfed resulted in a discussion about milk rationing. The academy adopted a formal appeal, recommending breastfeeding and the regulation of pasteurization and preservation of cow's milk. However, the real turning point occurred on 17 September 1940 when, in reaction to a proposal by Georges Duhamel,[16] the academy decided to create an 11-member Food Rationing Commission.[17] Its mission was to evaluate the rationing programme of the French population and to formulate predictions concerning its effect.

---

13   The minutes of weekly meetings are published in the *Bulletin de l'Académie de médecine (BAM)*.

14   Legislation had been passed in July 1938, but it order to avoid lowering morale, it was not implemented until spring 1940, Grenard 2007. See Chapter 13 by Kenneth Mouré.

15   Voeu de l'Académie de médecine 1940.

16   Duhamel had been a member of the Academy since 1937, despite the fact that he had ceased to practice medicine in order to pursue his literary interests.

17   Paul Le Noir (general medicine section), president, Gustave Roussy (general medicine section), Edmond Lesné (health and hygiene section), Francis Rathery (general medicine section, died in October 1941), Robert Debré (health and hygiene section), André Mayer (biological sciences section), Louis Tanon (health and hygiene section), Georges Duhamel (general member), Alphonse Baudoin (biological sciences section), Gabriel Bertrand (pharmacy section) et Pierre Martel (veterinary medicine section). The make-up of this commission would later change.

The academy was clearly not involved in drafting the rationing programme,[18] which had been called for by the occupying German government. Nevertheless, its members hoped to obtain dietary improvement for certain categories of the population[19] which it considered to be vulnerable to undernourishment, viz. infants, adolescents, manual labourers, pregnant women and nursing mothers. Another category were medical personnel for whom they requested T(ravailleurs) rationing cards. The academy seized on the opportunity of this 'favourable' context to promote nutrition education such as advocating a balanced diet.[20] On 8 October 1940, Edmond Lesné made the following declaration in the name of the academy,

> The current food shortages have made it necessary to ration foodstuffs; however, the strict restrictions (approximately 1,220 calories) and the deficiencies which are being imposed on the population could very well have a negative effect on the present and future health of French youth. For this reason, the Academy considers it its duty to inform the public powers of the need to distribute food according to the physical condition of each subject and to give doctors, social workers and housewives advice that will help them to compensate for the consequences of this undernourishment.[21]

Further, he recommended that doctors emphasize the benefits of breastfeeding,[22] that the diet of infants who were not breastfed be mixed from the age of three months onwards. He also suggested to carefully and regularly monitor the state of health of children and adolescents who were at risk of stunting and of nutritional disorders which could predispose them to tuberculosis. Housewives were also called upon not to waste foodstuffs.

At the end of the report, academy members made the following appeal to the authorities: children under one year of age who were not breastfed should have a minimum daily sugar ration of 30 grams; pregnant and nursing women, along with children and adolescents under 20 years of age, should receive double rations of meat, cheese and fat, the population should receive substitute provisions (fish, blood sausage, casein, soy flour, dried vegetables, vegetable oils) and the stock

---

18   Edmond Lesné, a member of the academy, chaired the consultative committee for provisioning.

19   The population was divided into different categories (J, A, T, C, V) which gave them access to different rations.

20   The war years were characterized by a mobilization of nutritionists, who were frequently called on by the government. The Société de diététique (Dietary Society), established in 1941, brought together the principal French nutrition experts. The debate about the consequences of under nourishment was a prominent topic in the society's journal, the *Revue française de diététique*, founded in 1941.

21   Lesné 1940: 652.

22   Equally recommended by the National Committee for Children of which Lesné was president from 1943.

of synthetic vitamins should be increased.[23] However, while the academy insisted on the need to limit the effects of rationing on the state of public health, they did not question the need for strict rationing. On 15 October 1940, Francis Rathery declared that, the 'Academy does not criticize the need for dietary restrictions . nor their overall caloric content. The hard truth that we must accept is that each French citizen will be subjected to a reduction in caloric intake'.[24] He concluded this statement with a reminder that while the official ration only covered 50 per cent of energy requirements, people had the possibility of supplementing their diet with unrationed foodstuffs such as potatoes, vegetables, eggs, fish, horse meat, rabbit, tripe products and blood sausage.

Members of the academy were at that point confident that they would be able to have an influence on rationing policy and thus function as a sort of privy council. They limited their actions to alerting the authorities to the consequences to be expected from food shortages and rationing, in the hope that their strategy would be successful. This was, however, illusory because ration levels had to be authorized by the occupying authorities, which regularly demanded reductions in the rations. The academy, therefore, in some way participated in the administration of the shortages.[25] On 28 January 1941, Rathery maintained that,

> We have explained the need to modify the food rations for the group of 'ill subjects' in order to make their medical treatment possible. By requiring both diseased and healthy inhabitants to follow the same caloric restrictions, we felt we had remained faithful to the very spirit of the government decrees, while at the same time allowing doctors to attempt to care for their patients. We are pleased to see that the authorities have adopted our guidelines, thus demonstrating their desire to attenuate the consequences of the dietary restrictions.[26]

Although academy members had an independent status (they were designated by their peers), they nonetheless were required to maintain total neutrality. In a political context that was notably marked by a ban on doctors' unions, academy members chose to remain discreet, in the hope that they could play a greater role in defining the new public health policy. In a speech made on 10 December 1940, Charles Achard, general secretary of the academy from 1920 to 1944, considered

---

23   The Academy noted that artificial vitamins could not compensate for an insufficient diet in March 1941.

24   Rathery 1940: 677.

25   They further warned against the dangers of intoxication from certain foods (rhubarb leaves, mushrooms, canned goods, food spoilage) and advised against intense athletic activity.

26   Rathery 1941: 80. Rathery referred to a decree dated 21 December 1940 which allowed for additional quantities of fats and meats for certain patient categories, upon presentation of a medical prescription.

it useful to recall the importance of that role.[27] However, we can equally postulate that academy members, along with the majority of the French population, were in a state of shock over the capitulation and thus partially adhered to Marshal Pétain's sacrificial rhetoric.[28] Food restrictions were part of the price that France had to pay in order to rise up again. Furthermore, in line with the majority of the medical profession, they placed great hopes in a 'national revolution' which would include an ambitious health policy that the academy had been advocating for years, but which had not been implemented under the Third Republic. On 10 December 1940, Achard further stated that, 'The role of public medical professionals in the reorganisation of our unfortunate country is in no way insignificant for the nation'.[29] This did not prevent the academy from congratulating the government for its determination in the struggle against alcoholism in July 1940.[30] The academy, further, approved projects to develop government-controlled medical services and sanitary education, which until that time had met with serious opposition from members of the medical unions who were strongly attached to the notion of private medical practice.

Accusations against the Vichy Regime in the Face of Rising Food Scarcity

It is difficult to establish a precise date for the academy's change in attitude towards the regime and, as the change was gradual and limited, it is equally difficult to speak of a veritable turnaround. The increasing difficulties in obtaining provisions were a determining factor. They can be attributed in part to the intensified effort by the German occupiers to drain the country's industrial and agricultural resources in support of the German war effort.[31] However, the scarcity of provisions was equally due to the Vichy collaborationist policy which began in autumn 1940 and was intensified by Pierre Laval beginning in April 1942.

As Francis Rathery had predicted, non-rationed foodstuffs – including food substitutes developed by researchers[32] – were becoming increasingly scarce from 1941 onwards. At the same time, more and more products were rationed and the nutritional value of the rations declined.[33] There was also a deterioration in food

---

27   In December 1940, Achard praised the academy's accomplishments in hygiene, in the fight against contagious diseases and infant mortality.

28   Laborie 1990.

29   Achard 1940: 878. Along with the majority of doctors, members of the academy considered that health conditions were a partial cause for the country's collapse.

30   The medical profession readily supported the anti-alcohol legislation of August 1940 and September 1941, Boninchi 2005.

31   Other factors were a reduction in agricultural production, farmers' reticence to deliver crops to the authorities and the black market, Grenard 2007.

32   For example, peanut oil meal, used to feed cattle in times of scarcity of other fodder, became as scarce as the meal that it was intended to replace.

33   Cépède 1961.

quality, particularly with regard to bread.[34] On 21 April 1941 French resistance member Liliane Schroeder noted in her personal diary, 'We managed to finish the week without anyone dying of hunger'.[35] Thanks to reports by doctors from hospital wards and private practice, the academy was able to record growing signs of nutritional deficiencies in their regular meetings. These were caused by the insufficient and poorly-balanced rations, which fell short in fats and protein. This led to increasing and significant weight loss, famine oedema, a rising number of cases of cachexia, growth retardation, an increased number of cases of tuberculosis and other infectious diseases and, finally, the spread of mycoses (scabies). However, members of the academy also became aware of their inability to act in this increasingly disturbing situation. Even if the sanitary authorities had accepted increased provisions for certain categories of the population,[36] they had less and less leeway and thus remained deaf to the academy's requests and warnings. The quantity of rations available with ration cards was regularly decreasing.[37] In hospitals and private practices it was becoming more and more difficult to obtain authorizations to feed up patients who were ill.[38] Well-informed observers were well aware of the meaningless nature of the recommendations drawn up by the academy. In his personal diary entitled *J'ai faim!* [I'm hungry!], Louis Le François wrote on 22 February 1942,

> I had a 'national coffee' with Doctor U. He said that the Academy of Medicine was studying the problems resulting from the food rationing programme. An Academy member presented a report on the nutritional value of skim milk and on the use of beef blood and horse meat. Another member gave a talk on individual and collective hysteria caused by rationing. I asked them: and what have you concluded? The answer: 'Human beings are incorrigible and refuse to live without eating'.[39]

In fact, despite the circumstances, academy members adjusted to the situation and continued to pursue their scientific activities. Extensive undernourishment gave them an unprecedented experimental framework for studying nutrition and diseases related to dietary deficiencies. Nonetheless, this statement of fact is overly simplistic. It is true that academy members were concerned about contributing to knowledge in their field, yet they also decided to take a stand. Their position can be observed in a certain number of papers and reports presented during the academy's

---

34   The Academy of Medicine obtained an increase in the extraction rate for bread flour.

35   Schroeder 2000: 75.

36   The most important was the establishment of a new population category – (J3) i.e. youth from 12 to 21 years of age – in June 1941.

37   All the more so because many rations remained unfilled due to a lack of goods.

38   Carnot, 30 March 1943.

39   Le François (pseudonym) 1942: 163.

meetings as well as in pressing appeals to the authorities from autumn of 1942 onwards. On 13 October 1942, Paul Carnot and his colleague Noël Fiessinger[40] proposed that the academy adopt an appeal calling for the allocation of a daily bread ration of 100 g for a three month period, renewable if necessary, for 'all adults in the A and V Categories who weighed more than 10 kilograms less than the weight recommended in the Quételet rule ... in order to improve the nutrition of all subjects who were at that very moment in serious danger of malnutrition'.[41] Several weeks later, on 17 November 1942, Paul Le Noir, the acting president of the Food Rationing Commission, presented an alarming report on 'the state of food rationing at the beginning of winter 1942–1943' in which he denounced the academy's fruitless efforts and described the state of undernourishment of the French population as a 'slow famine'.[42] As a result of this report the following appeal was adopted;

> The Academy of Medicine, concerned about the progressive reduction in the nutritional value of rationed foodstuffs available to the population, and equally troubled that it is becoming increasingly difficult, if not almost impossible, for the less fortunate to meet their minimum nutritional needs, believes it is its urgent duty to alert the French government to the public health risks that a prolonged state of undernourishment can create.[43]

These recurrent appeals seemed more and more unrealistic in the context of an increasingly severe nutrition crisis. In fact, it was the scarcity of foodstuffs – not a person's medical state – which was the determining factor in the quantity and energy value of rations received.[44] At the end of 1942, the government was still allocating food supplements to specific groups,[45] but the situation for most consumers continued to deteriorate. In the following months, the academy's position became more radical. On 9 February 1943, after a speech made by Charles Richet and Louis Justin-Besançon on 'the minimum needs of fat content in the human diet', a new firmer appeal was issued;

> Upon observing the progressive reduction in fats included in the rationing cards and disturbed that it has become almost impossible for the lower classes to obtain fats on the free market. ... [T]he Academy of Medicine believes it is once

---

40   This speech was the continuation of a talk presented a week earlier by Noël Fiessinger, entitled 'Current Weight Loss'.

41   Carnot 1942: 425.

42   Le Noir 1942: 497.

43   Le Noir 1942: 498.

44   This was a contributory factor in Jacques Leroy-Ladurie resignation from his post as Minister of Agriculture and Provisioning in September 1942.

45   These were pregnant women, prisoners, those with tuberculosis and patients in psychiatric hospitals.

again imperative to alert the authorities to the danger created by this inadequacy of fat content, be it animal or vegetable, and to the imbalance it can provoke in the human diet. This quantity of fats is much lower than the subsistence level. Should this deficiency persist or even more importantly, should it get worse, the number the deaths directly or indirectly related to the food restrictions would surely rise.[46]

The argument of an epidemic threat, which had been put forward in other appeals and reports, was henceforth systematically brandished. With memory of the 1918–19 flu epidemic still vivid, a typhus epidemic in the prisons of Marseille and Lyon in 1942, as well as an increasing number of rapidly fatal tuberculosis infections, were sources of great concern to health authorities.[47] With the start of the forced labour draft (le STO) on 20 April 1943, the academy adopted an appeal to the French population intended for the Vichy government as well as for the German occupier, 'The Academy of Medicine urgently calls the attention of the authorities to the dreadful risks which currently threaten the French population (particularly the threat of tuberculosis). The risk of increased, inevitable contamination equally extends to other foreign countries which employ workers from our country.'[48] This appeal was sent to Marshal Pétain, who responded in the following terms;

> Mr President, in a letter dated 22 April, you chose to inform me of an appeal of the Academy of Medicine, informing the authorities of the dreadful risks to the French population caused by a highly insufficient dietary regime. ... I beg to inform you that I am clearly aware of this serious matter. I will nonetheless take advantage of the Academy's appeal to once again inform the Head of the Government of this problem and to ask him to study any measures which might be taken to limit the serious risks that undernourishment of the French population may constitute for the future of the race.[49]

The tone of this letter contrasts with the compassionate image of Marshal Pétain conveyed in propaganda, particularly in posters printed by the National Aid Agency (*le Secours national*). The head of state, visibly irritated by the disrespectful attitude of the academy, turned down their request point-blank; his response can in some respects be interpreted as an admission of powerlessness.

Although the academy members' revolt can be explained by their concern about the deteriorating state of public health, other factors must also be considered.

---

46   Richet and Justin-Besançon 1943: 87. The Academy never mentioned the black market.

47   In *L'honneur de vivre*, published in 1974, Robert Debré declared that the doctors took advantage of the fear of epidemics to rebel against the German occupation.

48   Vincent 1943: 243.

49   This letter was published in the minutes of the meeting on 6 July 1943. *BAM*, 127: 382.

Firstly, the context of ever-increasing discontent; although the general public had for a long time remained attached to Marshal Pétain, there was in fact growing criticism of the regime.[50] This escalation of opposition can be largely explained by the difficulties encountered in obtaining provisions,[51] as shown, for example, by the increased number of protests by housewives. These difficulties were exploited by the resistance movements, which started to mobilize a growing number of doctors from 1943 onwards.[52] In an issue of the clandestine newspaper *Combat medical*,[53] the following statement can be found in an article entitled 'Racism? No! Defence of the people of France';

> The French race, or more precisely, the people of France, is certainly at great risk. Even those members of the French population who are less informed about notions of health and biology can clearly see that their conditions as individuals, in their families and as a nation, have been seriously affected by the living conditions to which they are currently exposed. This is all the more reason for doctors to be concerned about the deterioration of the general sanitary situation.

The author then quotes a statement made by academy member Victor Balthazard on 5 January 1943,

> with the current allocation of 1,200 to 1,400 calories, our dietary situation has become critical – the harshest of any time in recent history. ... It is therefore time to become alarmed about the future of our race and to fear the consequences of, for example, a flu epidemic, which could be exceptionally severe.[54]

The author of the unsigned article was probably Robert Debré, a member of the Food Rationing Commission. At the same time, he was in fact engaged in the resistance[55] with Louis Justin-Besançon[56] and Charles Richet, Jr.

---

50 Laborie 1990.

51 Other factors were the deportation of Jews, intensified repression and the STO (forced labour draft).

52 Under nourishment was equally denounced in the collaborationist press.

53 *Combat médical* was a publication of the National Movement against Racism, created in the spring of 1942. The principal objective of the article was to denounce the practice of tracking down Jews in hospitals and to encourage the Order of Medicine, the Society of Hospital Doctors and the Academy of Medicine to 'speak as a single voice' in opposing the practice.

54 Balthazard 1943: 6.

55 He went underground in September 1943 to join the medical committee of the resistance which was associated with the National Resistance Council (CNR). In October 1943, he became president of the Front national des médecins.

56 Another member of the Resistance Medical Committee, who became president of the French Red Cross after the Liberation.

Charles Richet, Jr. was elected to the academy on 29 October 1940 and he became the figurehead of the revolt. Richet was also a member of the very conservative CDL (*Ceux de la Libération*) resistance movement.[57] Arrested on 30 May 1943, he was deported to Buchenwald on 22 January 1944.[58] He was the instigator of the principal interventions regarding undernourishment, not only at the academy, where he gave numerous speeches and frequently participated in the debates, but equally in magazines such as *Paris medical*,[59] directed by Paul Carnot, or in the *Bulletin de la Société d'hygiène alimentaire*.[60] On 18 March 1943, on his own initiative, he addressed the medical profession in the *Bulletin de l'Ordre des médecins*. In his statement, he took the risk of predicting that two million French citizens were 'likely to succumb to famine, either directly or indirectly, as a result the development of infectious diseases'.[61] In his speeches to his students at the School of Medicine in Paris, his remarks were even more explicit. In a account given in 1995, Alexandre Minkowski describes how Richet would shout out 'The Krauts are starving the French population!' in the middle of his lectures.[62]

Conclusion

Under the pretext of fulfilling its duty as guardian of public health, the Food Rationing Commission, which included several notable opponents of the German occupier and/or the Vichy regime,[63] managed to overcome the excessive cautiousness of academy members and to convince them to denounce undernourishment frequently and explicitly. In doing so, members of the academy condemned both the German occupation as well as a policy which was jeopardizing the health and even the very lives of the French population. The academy was without doubt able to skilfully make use of its image as a somewhat marginalized institution whose expertise was challenged by newer, more modern organizations which were more dependent on

---

57   He was the head of its medical unit.

58   His colleagues, who knew nothing of his activities in the resistance, thought that he had been arrested for having virulently denounced the 'slow famine'. After the war, he became a specialist in deficiency diseases.

59   See the article 'Famine et épidémies', [Famine and epidemics] 10 September 1941, in which he speaks of plague, typhus, cholera, tuberculosis, smallpox, measles, typhoid fever and dysentery.

60   Richet 1941, Richet and Richet 1942. Gabriel Richet is the son of Charles Richet; he was equally engaged in the resistance and also deported.

61   Richet 1943: 59.

62   Archives de l'Académie de médecine: dossier C. Richet.

63   To be specific, Georges Duhamel, whose works were banned, or Gustave Roussy, dismissed from his post as director of education for the Paris academic sector because he did not put a stop to a student protest march on 11 November 1943. André Mayer, who was of Jewish background like Robert Debré, fled to Algeria in 1941 then to the United States in 1943.

the authorities[64] such as *l'Institut national d'hygiène*,[65] *la Fondation pour l'étude des problèmes humains*,[66] or even *l'Ordre des médecins*.

In this context, the impact of its appeals was probably quite limited. However, in choosing to dramatize a situation which could have degenerated and become a sanitary catastrophe, the first objective of the members of the Food Rationing Commission was to stimulate anti-German sentiment and to bring discredit upon a regime which was incapable of assuring the survival of the population. It further aimed to mobilize a 'wait and see' medical profession which, when confronted with the devastation of undernourishment, had seen its loyalty to Vichy seriously shaken.[67] This chapter, in any case, confirms the long-underestimated influence of shortages and dietary restrictions on the French population's growing disaffection with the Vichy regime.

References

Achard, C. 10 December 1940. La part de l'Académie de médecine dans la protection de la santé publique. [The role of the Academy of Medicine in the protection of public health] *Bulletin de l'Académie de médecine (BAM)*, 124, 878–90.

Alary, E. 2006. *Les Français au quotidien 1939–1949*. [French People in Everyday Life] Paris: Perrin.

Balthazard, V. 5 January 1943. Allocution. [Address] *BAM*, 127, 5–7.

Binet, L. 3 March 1942. Recherches fonctionnelles et biochimiques sur des personnes âgées. [Functional and biological research on the elderly] *BAM*, 126, 203–7.

Boninchi, M. 2005. *Vichy et l'Ordre moral*. [Vichy and the Moral Order] Paris: PUF.

von Bueltzingsloewen, I. ed. 2005. *Morts d'inanition: Famine et exclusions en France sous l'Occupation* [Death by Undernourishment: Starvation

---

64   The Vichy regime wanted to make up for France's slow progress in the field of epidemiology; numerous dietary studies were ordered by the inquiry commission of the Food Safety Society (Société d'hygiène alimentaire) in association with the National Hygiene Institute (INH) and the National Aid Agency (le Secours national). Vichy equally called on laboratories which specialized in nutrition research; for example, the laboratory of nutritional physiology of the Ecole des hautes études or the Dietary Research Centre of the Ministry of Agriculture.

65   Picard 2003.

66   Like the INH, the Fondation Carrel also had a nutrition research team. The Advisory Committee for Hygiene (Comité consultatif d'hygiène) equally had a dietary division and the National Aid Agency created a scientific dietary commission.

67   Other contributing factors include the obligation for doctors to break medical confidentiality and denounce wounded members of the resistance.

and Exclusions in France under the German Occupation]. Rennes: Presses Universitaires de Rennes.

von Bueltzingsloewen, I. 2007. *L'hécatombe des fous. La famine dans les hôpitaux psychiatriques français sous l'Occupation.* [The Sacrifice of the Insane: Starvation in French Mental Hospitals under the German Occupation] Paris: Aubier.

Bunle, H. 1947. La population de la France depuis 1939. [The French population since 1939] *Revue d'économie politique*, LVII, 816–63.

Carnot, P. 13 October 1942. Sur la communication de M. Noël Fiessinger relative à l'amaigrissement [About the report of Noël Fiessinger on weight loss], *BAM*, 126, 425.

Carnot, P. 30 March 1943. La sous-alimentation actuelle et ses conséquences digestives. [Current undernourishment and its digestive consequences] *Paris Médical*, 125, 85–9.

Cépède, M. 1961. *Agriculture et alimentation en France pendant la seconde guerre mondiale.* [Agriculture and Diet in France during the Second World War] Mayenne: Génin.

Chapireau, F. 2009. La mortalité des malades mentaux hospitalisés en France pendant la deuxième guerre mondiale : étude démographique. [Mortality of the insane in French mental hospitals during the Second World War] *L'Encéphale*, 35, 121–8.

Chevallier, A., and Moine, M. 23 January 1945. Sur l'évolution de la mortalité en France pendant l'Occupation. [The progression of mortality in France under the German occupation] *BAM*, 129, 44–9.

Fiessinger, N. 6 October 1942. L'amaigrissement actuel. [Current weight loss] *BAM*, 126, 422–6.

Grenard, F. 2007. Les implications politiques du ravitaillement en France sous l'Occupation. [Political implications of supplying provisions in France under the German occupation] *Vingtième Siècle*, 94, 199–215.

Jackson, J. 2004. *La France sous l'Occupation 1940–1944.* [France under the German Occupation] Paris: Flammarion.

Laborie, P. 1990. *L'opinion publique sous Vichy. Les Français et la crise d'identité nationale 1936–1944.* [French Public Opinion under the Vichy Regime: The French and the National Identity Crisis of 1936–1944] Paris: Seuil.

Le François, L. 1942. *J'ai faim.! Journal d'un Français en France depuis l'Armistice.* [I'm hungry! The Diary of a Frenchman in France since the Armistice] New York: Brentano's.

Le Noir, P. 14 October 1941. Voeu sur la pratique des sports et la sous-alimentation. [Appeal about the practice of sports and undernourishment] *BAM*, 125, 194.

Le Noir, P. 17 November 1942. Sur l'état du rationnement alimentaire au début de l'hiver 1942–1943. [The state of food rationing at the beginning of winter 1942–1943] *BAM*, 126, 496–500.

Lesné, E. 8 October 1940. Sur le rationnement alimentaire. [About food rationing] *BAM*, 124, 652–66.

Picard, J.F. 2003. Aux origines de l'Inserm: André Chevallier et l'Institut national d'hygiène. [The origins of the National Institute of Health and Medical Research: André Chevallier and the National Hygiene Institute] *Sciences sociales et santé*, 21(1), 5–26.

Rathery, F. 15 October 1940. Rapport sur la ration du sujet malade. [Report on rations needed for the ill] *BAM*, 124, 677–83.

Rathery, F. 28 January 1941. Le personnel médical devant les restrictions alimentaires. [The medical staff confronted to food shortages] *BAM*, 125, 80–90.

Richet, C. 1941. Etudes sur le rationnement alimentaire à l'Académie de médecine. [Studies on food rationing presented at the Academy of Medicine] *Bulletin de la Société scientifique d'hygiène alimentaire*, XXIX, 72–9, 109–16.

Richet, C. Mai 1943. Aux médecins de campagne. [To country doctors] *Bulletin de l'Ordre des Médecins*, 3, 59.

Richet, C. and Richet, G. 30 October 1942. L'alimentation actuelle des Parisiens. [The current diet of Parisians] *Paris médical*, 43, 226–7.

Richet, C., Lesueur, G., and Duhamel, G. 5 January 1943. Formes irréductibles de l'insuffisance alimentaire chez l'adulte. [Irreducible forms of undernourishment in adults] *BAM*, 127, 7–10.

Richet, C. and Justin-Besançon, L. 9 February 1943. Sur le besoin minimum de graisses de l'alimentation humaine. [Minimal fat needs in the human diet] *BAM*, 127, 84–7.

Rollet, C. and De Luca, V. 2005. La vulnérabilité des enfants: Les crises de mortalité de 1940 et 1945, [The vulnerability of children: the mortality crisis of 1940 and 1945] in *Morts d'inanition: Famine et exclusions en France sous l'Occupation* [Death by Undernourishment: Starvation and Exclusions in France under the German Occupation], edited by I. von Bueltzingsloewen. Rennes: Presses Universitaires de Rennes, 263–79.

Schroeder, L. 2000. *Journal d'Occupation: Paris 1940–1944.* [Occupation Diary: Paris] Paris: F.X. de Guibert.

Veillon, D. 1995. *Vivre et survivre en France 1939–1947.* [Living and Surviving in France] Paris: Payot.

Vincent H. 20 April 1943. Voeu relatif à l'alimentation de la population française. [Appeal about the feeding of the French population] *BAM*, 127, 242–3.

Vincent, P. 1946. Conséquences de six années de guerre sur la population française. [The consequences of six years of war on the French population] *Population*, 3, 429–40.

Voeu de l'Académie de médecine [Appeal of the Academy of Medicine], 28 Mai 1940, *BAM*, 124, 436.

Chapter 13

Réalités cruelles: State Controls and the Black Market for Food in Occupied France

Kenneth Mouré

Introduction

Food rationing was a 'painful necessity', Marshal Philippe Pétain told French citizens in October 1940. 'We have not tried to avoid cruel realities and, faced with the need to impose harsh restrictions, we wanted to ensure that all would sacrifice equally. Each must share in the common privations, such that wealth does not save some and increase the misery of others.'[1] The Ministry of Agriculture announced that rationing would avoid 'intolerable injustices'; 'the use of ration cards will assure the same rights to everyone regardless of their wealth'.[2] This egalitarian imperative and a period of necessary sacrifice would forge a new France inspired by common ideals and dedicated to the national interest.

Realities were cruel indeed. Hunger triggered desperation rather than a spirit of sacrifice, and the easy profits from illicit trade proved far more alluring than commitment to the state's new values of work, family and country. The rationing system under Vichy did not provide equality in the face of sacrifice. Rather, it promoted evasion, 'parallel' markets, and the individualistic 'spirit of lucre and of speculation' that Pétain claimed Vichy's 'coordinated economy' would bring to an end.[3] The state-administered food economy suffered escalating losses to a growing black market, where high prices and high profits attracted output and trade from official markets. The demand for scarce goods and the premium for risk priced many black market goods beyond reach for those of modest means. Inequities in food supply deepened social divisions and animosities. Between the extremes of official markets bare of goods and thriving black markets charging exorbitant prices, a range of other possibilities developed to supply food. Consumers survived the shortages through a combination of expedients. The first two winters of the occupation were the most difficult; new consumption strategies developed to alleviate the impact of insufficient rations until the Allied invasion disrupted supply and transport in 1944.

---

1 Pétain 1989 (10 October 1940): 84.

2 Archives Nationales, Paris (AN), F 60 1546, Communiqué of 29 September 1940.

3 Pétain 1989: 92. British efforts at 'fair shares' for all were much more successful, see Chapter 10 by Ina Zweiniger-Bargielowska.

French experience with rationing under the occupation demonstrates, on the one hand, a system unable to provide sufficient food. On the other, a combination of market and consumer responses made up for some shortages, but in doing so *increased* the defects in the rationing system. After a brief overview of the extent of and reasons for the food shortages in wartime France, this chapter addresses: first, the organization of the food rationing system; second, the development of alternative distribution networks, and third, consumer strategies to obtain food for survival. At the heart of these three concerns is the relationship between state controls to manage scarcity and market responses to shortages and controls. The concluding section assesses the relationship between shortages, controls and food consumption in occupied France.

Food Shortages

Pétain promised that his government's first priority would be to assure 'sufficient food for all' and that the rationing would share sacrifice in response to the immediate impact of military defeat, transport and population chaos and German demands. French plans assumed this was a temporary crisis and that restoring order to the food supply system and further negotiation would assure sufficient food for all. In fact, this began a four-year period of decline. The immediate problems of disrupted production and transport, and the longer-term problem of the shortage of manpower, were aggravated by large-scale German seizures, purchases and requisitions; the Allied blockade; the progressive losses of vehicles, draft animals, fuel and fertilizer; the closing of Mediterranean trade in November 1942; and finally, renewed disruption of the harvest and food transport in 1944.

By 1944, French agricultural production had fallen by about one third.[4] German requisitions and direct purchases were taken from this reduced supply. The reductions from prewar supply levels were 49 per cent for sugar, about 50 per cent for potatoes (annual Parisian consumption prewar had been 165 kg per person; in 1943 it was 50 kg). The decline with regard to butter was 55 per cent, 95 per cent for oil (loss of imports) and 60 per cent for milk (with increased farm consumption). Grains for bread fell by 26 per cent. Average meat consumption prewar, not including on farms, had been 596 g per week, compared with rations of 90 g per week in 1944 (often unfilled). For most of these goods France had been nearly self-sufficient. Imports declined sharply and the French consumed only 11 per cent of their prewar coffee consumption during the occupation.[5] Food

---

4   Cépède 1961: 324, there was a decline of about 20 per cent in grain and vegetable output, and of about 40 per cent in meat output.

5   AN, F 37 120, These statistics are drawn from reports produced by the Commission Consultative des Dommages et des Réparations in 1946–47.

imports fell to barely 5 per cent of their 1937–38 volume by 1943–44.[6] At the level of national aggregates, food available to French citizens in 1944 was likely little more than half that of 1938, a dramatic contraction from relative abundance to bare subsistence for many.

The impact of shortages varied greatly by region and by consumer income. The worst-hit regions were those with a dry climate and regional monoculture such as the Hérault, Alpes-Maratimes and Jura. These particularly barren regions relied on food from other regions and they were, therefore, acutely vulnerable to shortages and transport disruption.[7] Urban centres suffered greater hardship than rural areas with gardens and diverse agricultural output. Residents of cities in rich agricultural regions could draw on regional produce, black markets, weekend foraging trips to the countryside and could arrange for family parcels. In rural areas with diverse agriculture, farm consumption increased significantly and profits from grey and black market sales improved both health and wealth. The mortality rate in France was 11.6 per cent higher in 1941–43 than in 1936–38, with the greatest increase in urban departments. There were increases of 57 per cent in the Bouches-du-Rhône (Marseille), 29 per cent in the Rhône (Lyons) and 24 per cent in the Seine and Seine-et-Oise (Paris). Mortality rates *fell* by more than 10 per cent in the Indre, Mayenne and Orne; departments in Normandy and Brittany also fared well.[8]

The Rationing System

The rationing system began amidst chaos. The flight from advancing German armies had strained local supplies in departments with refugees.[9] Local shopkeepers, mayors and prefects improvized rationing measures to conserve goods and keep order in queues. Some prefects banned the export of food from their departments. Throughout the summer, transport difficulties and the German division of France into separate zones hindered national distribution. At the same time, German requisitions and massive purchasing combined to create serious regional shortages.[10] The French

---

6 Food imports fell precipitously, from an index value of 100 in 1937–38, the tonnage of food imports was 47.5 in 1941, 27.3 in 1942, and under 6 in 1943 and 1944. INSEE 1950: 253.

7 The *Cahiers de l'Institut d'histoire du temps présent* survey of regional inequalities during the 'time of restrictions' divides departmental studies into three categories: 'nourriciers' [feeding others], 'affamés' [starving] and 'intermédiaire' [intermediate] – the starving departments include the Hérault, the Apes-Maratimes and the Jura. Veillon and Flonneau 1996.

8 Cépède 1961: 403–7.

9 Diamond 2007: 74–9; for one prefect report on 'the almost insurmountable difficulties' of trying to feed tens of thousands of refugees see AN, AJ 41 377, report from Department of the Marne, 1 Oct. 1940.

10 AN, F 60 1546, see 'Note sur le ravitaillement de la population civile', 3 August 1940. For a local report on problems in Nantes, see the notes from P. Brossier 16 August to

authorities wished to restore order and regain control of stocks seized by the *Wehrmacht* [German army]. In early July, Intendant General Casanoue sought authority to organize food distribution, provide manpower for the grain harvest, restore transport across the demarcation line and recover food supplies seized by the *Wehrmacht*.[11] The German authorities demanded that France impose a comprehensive national plan for food rationing. They also stipulated oversight of all administrative appointments and policy.[12] The Germans calibrated French rations to ensure that French consumers were distinctly less well off than civilians in Germany. A 'unilateral and irrevocable decision of the German High Command' set French rations at 350 grams of bread per day, 350 grams of meat and 80 grams of fat per week, and 500 grams of sugar per month.[13] German civilians in France would receive German rations; the French would have rations comparable to those in Germany in the worst period of the First World War.[14] French officials believed the Germans wished to impose conditions equivalent to the worst period of the First World War in Germany.[15] French citizens recognized the need for rationing, but dismayed at the low level for rations, hoped for measures to increase rations, especially for bread and meat.[16]

Pétain's promises of sufficient food and shared sacrifice assumed that the situation would improve and that France would be able to negotiate increased rations. Neither assumption held. Bad weather, blockade, shortages of manpower, fuel and fertilizer and increasing German requisitions destroyed most possibilities for increased rations. Concessions gained in negotiations on food supply were rare and at best marginal.[17] The trend for rations was downward, with caloric value falling from 1,250 per day to 1,040 for adults who actually received their full rations.[18] German policy sought to maximize resources extracted from France. Any changes in quantity or price required German approval and, therefore, rapid adaptation to changes in local supply was impossible. German authorities adamantly opposed raising prices, although prices fixed too low to bring goods to market were an obvious problem for German as well as French supplies. In negotiations the Germans blamed the French for any problems, denied sensible French proposals, and delayed decision on others before rejecting them. The Germans paid no heed to international conventions or the armistice agreement they had signed in June 1940; France would contribute

---

4 October 1940, Bourderon 1970.

11   AN, F 60 1547, Casanoue to Reinhardt, 7 July 1940.

12   AN, F 60 1547, Reinhardt to Casanoue, 11 July 1940.

13   AN, F 60 1546, Général de Corps d'Armée to Pétain, 'Plan de rationnement alimentaire', 8 September 1940.

14   AN, F 60 1546, 'Plan du rationnement alimentaire', 8 September 1940.

15   AN, F 60 1946, Schoppmann to Casanoue, 28 September 1940, Rist 1983: 94. German residents in France would be allowed the same rations as if they were in Germany, allowing about 2,570 calories per day for adults.

16   Archives de la Préfecture de Police, Paris, 'Situation à Paris', 23 September 1940 and following weeks.

17   Cépède 1961: 154–60, Boussard 1974.

18   Cépède 1961: 386, 389.

food supplies to the German effort to defend Europe against Bolshevism.[19] When German food shortages reached a crisis point in 1942, France and Ukraine were to be the key sources for additional food. There were limits to what could be wrung from France without increasing civilian resistance and reducing productivity, as military authorities in Paris pointed out in protesting Berlin's demands for higher food requisitions in 1942.[20] But German policy deliberately strained the French food economy, at a cost to the health of French citizens and the productivity of French agriculture and industry.

France used controls on both prices and quantity to manage the distribution of scarce resources. Neither price controls nor rationing worked well. Both began in the *drôle de guerre* [phoney war] but were seriously applied under the adverse circumstances of dramatically reduced supply and transport, a disoriented and deprived population and a German Occupation bent on exploiting French resources. Food rationing, to distribute scarce food equitably, must provide sufficient food, the rations serving as a guaranteed minimum. Public confidence is essential to prevent the loss of supplies from official to black markets. The legitimacy of the rationing system is critical in determining whether production and distribution favour official markets or to alternative supply networks. In France, the inadequacy of rations was obvious and prompted the development of alternate markets. Black market activity began immediately to meet demand for scarce goods in the summer of 1940. The enforcement of price controls and rations received serious attention in the autumn of 1940, as illicit activity spread. Enforcement was never able to recover the head start gained by the black market.

The state *had to allow* citizens to obtain additional food, given the inadequate rations. Vichy could not sustain a system of deliberately starving its people. There were many means of obtaining extra food, legal and illegal, and a system of increasingly complex rules and inconsistent enforcement blurred the distinctions between the two. Most consumers followed a natural progression from seeking extra food, to not explicitly forbidden transactions, to black market practice. In state policy, three significant developments can be noted that opened opportunities for illegal supply. The first allowed sending food by parcel from countryside to city; these 'family parcels' were considered essential to supplement rations. The parcels began to be sent on an informal basis between family and friends in 1940. Formal authorizations for 'family parcels' in 1941 and 1942 facilitated the growth of a distribution system that provided a vital source of food for urban residents and for black market restaurants. Secondly, measures to feed various collectives gave priority to specific groups in the ration system. In each case – factory canteens, gardens in public spaces, and *restaurants communitaires* for low-income workers – privileged groups could draw food from the official food supply ahead of

---

19   AN, F 37 155, Casanoue, 'Note sur les prestations alimentaires et les livraisons de produits agricoles exigées par les autorités d'Occupation' [Note on the food services and the agricultural supplies required by the Occupation authorities], 20 March 1945.

20   Mitchell 2008: 64–6.

individual purchasers.[21] Finally, a relaxation of the rules for small-scale transport of goods was needed to assist families and to reduce the public animosity toward state regulations that seemed to benefit only the Germans. Such a relaxation had been urged and adopted by department prefects; Secretary General of Police René Bousquet made it official in July 1942, writing to prefects;

> We need to put an end to annoying interventions and to clearly excessive penalties, to which consumers have been too often subjected when they are simply trying to feed their families. Any measure ... that would irritate public opinion, which is already upset by the current difficulties in food supply, must be rigorously prohibited.[22]

All three measures reduced food supply in the official rationing system. The Germans objected to family parcels as a loss of food available for requisitions. The public complained repeatedly that control agencies picked on the small fry, leaving the big black market operators at liberty. By 1943 there was frequent complaint that the collectivities with priority in food purchasing left little or nothing for individual consumers. Privileges fostered their own patterns of abuse in each case. Family parcels and purchasing for collectivities both became important conduits for black market traffic. Relaxing enforcement of transactions to meet family needs blurred the boundary between legal and illegal supply, creating an ill-defined grey area of permissible infractions. Economic violations were widespread. Responsibility for controlling and containing them was divided among several enforcement agencies, all understaffed and poorly equipped. This produced inconsistent enforcement, further emphasizing systemic inequities.

German behaviour was still more problematic. The Germans had imposed the rationing system and ration levels that made the parallel economy necessary. They demanded food requisitions far beyond the needs of their occupation troops. German purchasing agencies set up offices in Paris and German armed forces used their massive purchasing power to supply troops and send a vast array of goods for shipment to Germany. This opened up extensive networks for black market purchases in both the occupied and unoccupied zones. The traffickers and middlemen employed were mostly French, but the Germans provided them with vehicles, fuel, passes and funds to operate throughout France.[23] German authorities intervened when traffickers working for them were arrested by the French police (whom the Germans exhorted to crack down on the black market), releasing the traffickers and their goods. For the Germans, black market purchases were a means to further exploit the French economy. For the rationing system and the

---

21　See Chapter 11 by Peter Atkins on British Restaurants, which provided much better fare.

22　Centre des archives économiques et financiers, Savigny-le-Temple [CAEF], B 16041, Director in Limoges, report of 1952.

23　Sanders 2001.

enforcement of controls, German actions drained resources from official markets and fostered the networks of supply and the desire for financial gain that structured and promoted the black market.

Alternative Distribution Networks

A number of legal alternatives were available to supplement rations, including purchasing unrationed goods, bartering, growing one's own food, obtaining goods from friends and family in the countryside and direct purchase of goods from producers. A further range of alternatives broke the law, such as sales of quantities or at prices in excess of those allowed, the transport of goods without permits, ration card fraud and theft. All options were exercised, but some proved more durable than others. Unrationed goods tended to rise rapidly in price and to vanish from official markets when their prices were fixed or when subject to rationing. Growing one's own food required access to land; where one lived (in the countryside, the suburbs or the inner city) often determined whether this was possible. Municipalities opened vacant terrain for gardens, and some businesses established gardens to grow food for factory canteens and distribution through workers' co-operatives.

'Family parcels' were the most frequent and important supplement. Edmond Dubois claimed that virtually everyone in Paris either had relatives or friends in the countryside, or employed servants who did.[24] Their use started in 1940 and in 1941 the state set rules specifying quantities and goods that could be sent in this way.[25] In December 1941, more than three million parcels were sent into the Department of the Seine by post.[26] One observer estimated in 1943 that more than 350,000 tons of food was sent into the Paris region, with at least as much again carried in by individuals for family consumption.[27] Those without relatives in rural areas sought friends and relatives of friends. The writer Colette received a steady supply of food from two admiring fans who were farming in Normandy.[28] German authorities complained that this diverted goods from official markets and reduced food available to them, but French officials persuaded them that ending the parcels would further increase black market activity.[29]

The black market for dairy products and meat was substantial. The illicit slaughter of livestock was widespread; by one estimate, every butcher in France was guilty of the practice. The black market was thought to be at least equal to the official market for volume of meat distributed in 1943. It was so widespread

---

24   Dubois 1946: 129–30.
25   Veillon 1995.
26   Sauvy in *Bulletin Brique rouge* no. 11, July 1942, 13.
27   Debû-Bridel 1947: 45–6, Veillon 1995: 173–6.
28   Colette 1992.
29   CAEF, B 16039, Leménager, 'Étude sur le contrôle des prix', January 1953, 586–7.

that it proved impossible to confiscate illicit profits from clandestine slaughter after the war.[30] Local communities protected farmers and butchers who had served community needs. Mayors were required to report the slaughter of animals, pigs especially; 'But they have no clue what's going on,' Henri Drouot recorded in Dijon; 'since the war, the pigs having their throats cut no longer squeal ...'[31] The meat supply fell precipitously. Rations were reduced to 90 grams per week and they often went unfilled for weeks in major cities. Pétain claimed in June 1943 that 350,000 tons of meat were sold on the black market and no one disputed this. The food supply service thought this understated the problem.[32] Though butchers' shops were empty, in black market restaurants it was considered 'smart' in spring 1944 to eat two or three steaks at one time.[33]

Direct purchases from farms were ubiquitous. They varied widely, from massive German purchases transporting agricultural produce in army trucks and black market operations by urban *trafiquants* selling to German purchasers with German-authorized vehicles, fuel and permits, to small-scale operators carrying goods to urban centres. This was the most common form – individuals seeking food for themselves and their families. Small-scale purchasers often relied on public transit – trains and metro – and on bicycles. Urban purchasers returning on trains were an obvious target for controllers; tight controls caused public hostility and usually netted only small quantities of food transported by women and children.[34] Passengers avoided station controls in some cities by alighting in the suburbs.[35] Bicycle use was also important: Henri Drouot recorded that 50 to 75 per cent of the population of Dijon made direct purchases in the countryside, with tens of thousands doing so by bicycle each day.[36] In July 1944, with controls on individual purchases lifted in the supply crisis after the Normandy invasion, thousands of cyclists from Paris were buying an estimated 15 tonnes of butter a day in the region of Gournay-en-Bray (near Rouen) and carrying it back to Paris on their bicycles.[37]

Shopkeepers and restaurant owners had easy opportunities to combine licit and illicit sales. Customers in queues grumbled about the 'back door' commerce that reserved the best goods for clients willing to pay extra, particularly for meat. The best-known literary black market commerce, the Poissonards' dairy shop

---

30   Mouré and Grenard 2008.

31   Drouot 1998: 121, Limouzin 1983: 82–3.

32   AN, AJ 41 98, Laporte, 'Note au sujet de la ration de viande' [Note on the subject of the meat ration], 15 September 1943.

33   Vailland 1948: 39.

34   Taylor 2000: 122–5, commenting on controls in Lille, April 1942, prior to the relaxation of rules for feeding one's family.

35   Drouot 1998: 446.

36   Drouot 1998: 261, 274.

37   AN, F 7 14895, Chief of regional services, Comité Central des groupements interprofessionnels laitiers, to Intendant of Police, Préfecture de la Seine Inférieure, 17 July 1944.

in *Au Bon Beurre*, was an exaggerated scenario, but shopkeepers were widely termed '*les bofs*' for their illicit commerce in butter, eggs and cheese (*beurre, œufs et fromage*).[38] Food shops, markets and restaurants were the principal targets for economic controllers, who usually imposed fines. Rigorous repression met with public hostility and new forms of evasion. The director of economic controls warned in April 1944 that, 'We cannot pursue impossible goals in requiring the population to abstain completely from meat because our collection system is deficient'. Therefore, cracking down on meat sales by butchers would simply increase clandestine butchering in the countryside, where control was more difficult.[39] Closing shops caught in black market traffic was one option, but when consumers registered to collect rations at specific shops, closing them required finding new outlets for those consumers.[40]

Black market restaurants provided lavish fare for those who could afford it. Vichy implemented a strict system regulating meal composition, prices, menus, days scarce foods could be served, and the provision of ration tickets. Abuses were a constant concern, in part because German authorities insisted on the need to purge the black market for restaurant meals, in part because the more egregious black market restaurants fed a *clientèle* whose income sources were suspect. The Germans, again, were part of the problem. They exempted six Paris restaurants from all rules in order to provide officials and visitors with luxury cuisine. They patronized many other restaurants, as did black market profiteers. These restaurants had German protection from French authorities and access to black market supplies. Good black market restaurants took pride in serving a 'prewar menu'.[41] The range of restaurants benefiting from German patronage is evident in the long lists of restaurants closed by the police after the Liberation in 1944.[42] Police surveillance turned up regular and repeated infractions without seriously curbing this field of illicit commerce. Most restaurants bought food on the black market that they could not otherwise obtain and they served extras at higher prices to make up the cost.[43]

---

38   Dutourd 1952; the narrator repeatedly notes that 'les bofs' are growing ever richer in Calaferte 1993.

39   CAEF, B 57659, Sailly to Bonnefoy, 1 April 1944.

40   CAEF, 4 A 3, A. Lebelle, 'Note sur l'insuffisance de la répression en matière de contrôle des prix' [Note on the insufficiency of repression in matters of price control], 26 July 1941.

41   Vailland 1948: 54.

42   CAEF, B 57659, there is a series of closures listed in this carton.

43   CAEF, B 49516, Service de Contrôle Économique, Department of the Seine, 'Rapport mensuel du mois de décembre 1943' [Monthly report, December 1943].

Consumer Strategies

Most consumers could not afford black market prices and rarely purchased black market goods. Those constrained to live on official rations simply could not survive. Patients in mental asylums often had to live on rations alone. Famine-related deaths in asylums totaled nearly 45,000, mainly in the period 1940–42, a toll alleviated when special rations were established for patients in December 1942.[44] The elderly were vulnerable by their age and their isolation. Personal connections could make all the difference in tapping into alternate distribution networks. Linkages to family and friends in the countryside were particularly important for arranging family parcels and foraging trips. Children were sent to stay with family in rural areas, or signed up for rural *Chantiers de la jeunesse* to have access to more food.[45]

Queues for food, which could consume vast amounts of time for little return, became an important social lieu for making contacts and obtaining information.[46] 'Interminable queues to get potatoes or milk', Annie Vallotton recorded in September 1941; in May her sister shared queuing for nearly five hours to obtain 100 grams of sausage, two cans of sardines and a *faux roquefort*.[47] Families with children could share the task, joining multiple queues or taking turns in line. *Cartes prioritaires* gave mothers and war veterans priority in queues, and they could use this to supply others, whether veterans buying for friends on commission, or mothers buying for neighbours.[48] Those living alone were at a disadvantage; they could not queue and work at the same time. Those with sufficient wealth could avoid queues and have goods delivered to their door. For most city dwellers, obtaining sufficient food became their major preoccupation, far more important than political and military events that did not impact their daily lives directly.[49]

Controls provided opportunity for fraud to supplement inadequate rations. Ration cards and tickets were widely available on the black market, the price differentiating between stolen and counterfeit cards. The assignment of card categories varied; in some regions most adults were given T cards allowing higher rations for heavy labour. Holding more than one card multiplied the rations one could obtain. Strategies included the purchase of extra cards, filing for the replacement of cards allegedly 'lost', and keeping and renewing the cards for deceased relatives. Rural families sent cards for goods their farms produced to relatives in cities. The theft of ration cards from municipal offices reached epidemic proportions in

---

44 von Bueltzingsloewen 2007, see Chapter 12 by Isabelle on Bueltzingsloewen.
45 Ruffin 1979.
46 Mouré and Schwartz 2007: 281–4, Veillon 1995.
47 Vallotton 1995: 176, 166.
48 Mouré and Schwartz 2007: 268, Drouot 1998: 71.
49 Edith Thomas noted the gap between the drama of major war events and daily life narrowly focused on food, tickets and queues; Thomas 1995: 166. Prefect and police reports routinely reported food supply as the principal public concern.

1943.[50] Government officials benefited from buying privileges and multiple cards. Admirals in the French navy were entitled to hold 17 cards. Minister of Agriculture and Food Supply Jacques Le Roy Ladurie, as a parting sign of his knowledge of abuses, left Pierre Laval a list of all the official vehicles caught with black market goods during his tenure as minister.[51] The director of the staff buffet at the *Banque de France* was implicated in black market activity when economic controllers stopped a *Banque de France* truck, officially transporting currency notes to Paris; gendarmes found ten tons of sugar and 864 litres of brandy.[52]

State officials often sided with the local populations, seeking to keep goods out of German hands and to meet local needs. Peasants around Dijon gave good prices to gendarmes in exchange for tips on new control initiatives and surveillance.[53] In conditions of severe shortages and administered penury, officials in areas with adequate food shared an interest with producers and consumers in keeping food available for the local population.

Thefts were common in rural areas as well as cities. Frequent thefts from rural gardens were known as '*le ravitaillement sans ticket*'[food supply without ration tickets].[54] Goods in transit, particularly the family parcels shipped by rail and post, were a major temptation. Thefts from rail cars, rare before the war, reached epidemic proportions despite SNCF efforts to increase security. In 1942 there were 134,790 thefts, taking merchandise valued at more than 89 million francs; 18,334 people were arrested for theft and 2,893 SNCF employees dismissed.[55] Thefts from postal delivery, from parcels being shipped to prisoners of war, and from municipal offices, were widespread. Thefts from POW parcels were singled out for severe punishment by the *Tribunal d'État*.

The decline in ration quantities made supplements ever more essential. As opportunities to obtain legal supplements dwindled, the state blurred the distinction between licit and illicit practices. Consumers had difficulty negotiating the tangled web of changing rules. Dominique Jamet provides a comic description of a grocer's store with a confusion of cards, tickets and rules, 'No one understands anything, everyone argues, everyone cheats or tries to cheat.'[56] Irregularities in enforcement, the sympathy and complicity of local officials and the seeming abundance of black market goods made the benefits of stepping beyond the rules irresistible. In addition,

---

50   AN, AJ 41 430, in the first nine months of 1943 there were 543 attacks to steal ration cards, 495 on municipal offices. Darnand to regional prefects, 15 February 1944.

51   Le Roy Ladurie 1997: 392.

52   Archives of the Banque de France, 1060200101 71.

53   Drouot 1998: 289, 697.

54   Drouot 1998: 197, 239, 276, 289, 310, 318, 500, 549, 664, for the Limousin, see Fogg 2008: 51–4, 95–100.

55   AN, 72AJ 1927, 'Vols en cours d'expédition par fer' [Thefts of goods being shipped by rail], Mar. 11, 1943, and Directeur du Service Central du Mouvement to M. le Baron d'André, Mar. 1943.

56   Jamet 2000: 64–5, he explains his family relied on regular weekly parcels from countryside.

there was a growing belief that goods on the black market were goods kept from the Germans. Therefore, their diversion and purchase was an act of patriotism.

Conclusion

Hunger was the defining experience for most French citizens during the German Occupation. It did not reach famine proportions because France, rich agriculturally, had the capacity to feed itself. The occupation reduced output, limited transport and took a significant quantity of scarce food. The French rationing system was flawed from the outset, implemented with little planning, calibrated to meet German demands and subject to damaging interference. German control made the rational use of prices to manage supply and demand impossible. Maintaining low official prices encouraged French farmers to produce less, consume more, and sell goods directly to purchasers who would pay higher prices.[57] Inadequate rations impelled consumers to find alternate means of supply and forced the state to allow the infractions necessary to feed families. Structurally, the system created forces that would undermine its legitimacy and efficiency. The black market thrived.

For consumers, the shortages, the unreliability of rations, the complexity of regulations and the irregular nature of enforcement (officials were often both sympathetic and themselves engaged in similar grey transactions) made the question of legality less important and less clear. French reliance on the *système D* – improvisation to solve problems – has a central place in the lore of occupation experience. Food was the most critical area for improvisation. This fostered alternate systems of distribution, restoring the role of prices to stimulate supply and allocating scarce goods to those who could pay high prices. Rather than 'equal sacrifice' and social solidarity, the system encouraged black markets and social division. One result of this peculiar collaboration between German interference and deficient French planning was a reverse Robin Hood effect, with black markets in effect taking from official markets accessible to the poor in order to feed the rich.

References

Bourderon, R. 1970. Le ravitaillement et les prix dans le département du Gard (été-automne 1940). [Food supply and prices in the department of the Gard (summer-fall 1940)] *Revue d'histoire de la Deuxième Guerre mondiale*, 79, 37–60.

Boussard, I. 1974. Les négociations franco-allemandes sur les prélèvements agricoles: l'exemple du champagne. [Franco-German negotiation of

---

57   Limouzin states farmers in the Corrèze sold to friends, traded for other goods, and fed the needy, but refused to sell to black market buyers, Limouzin 1983: 46, 78–83.

agricultural requisitions: the example of champagne] *Revue d'histoire de la Deuxième Guerre mondiale* 95, 3–24.

von Bueltzingsloewen, I. 2007. *L'hécatombe des fous: la famine dans les hôpitaux psychiatriques français sous l'Occupation.* [The Killing of Mental Patients: The Famine in French Psychiatric Hospitals during the Occupation] Paris: Aubier.

Calaferte, L. 1993. *C'est la guerre.* [This is War] Paris: Gallimard.

Cépède, M. 1961. *Agriculture et Alimentation en France durant la IIe Guerre Mondiale.* [Agriculture and Food in France during the Second World War] Paris: Éditions M.-Th. Génin.

Colette. 1992. *Lettres aux petites fermières,* [Letters to the Lady Farmers] edited by M.-T. Colléaux-Shaurang. Paris, Le Castor Astral.

Debû-Bridel, J. 1947. *Histoire du marché noir (1939–1947).* [History of the Black Market] Paris: La Jeune Parc.

Diamond, H. 2007. *Fleeing Hitler: France 1940.* Oxford: Oxford University Press.

Drouot, H. 1998. *Notes d'un Dijonnais pendant l'Occupation allemande 1940–1944.* [Notes by a Dijon Resident during the German Occupation] Dijon: Editions universitaires de Dijon.

Dubois, E. 1946. *Paris sans lumière 1939–1945.* [Paris without Light] Lausanne: Librairie Payot.

Dutourd, J. 1952. *Au bon beurre.* [The Best Butter] Paris: Gallimard.

Fogg, S.L. 2008. *The Politics of Everyday Life in Vichy France.* New York: Cambridge University Press.

INSEE. 1950. *Mouvement économique en France de 1938 à 1948.* [Economic Change in France from 1938 to 1948] Paris: Imprimerie nationale.

Jamet, D. 2000. *Un petit parisien 1941–1945.* [A little Parisian] Paris: Flammarion.

Le Roy Ladurie, J. 1997. *Mémoires 1902–1945.* [Memoirs 1902–1945] Paris, Flammarion.

Limouzin, R. 1983. *Le temps des J 3.* [The time of J 3s] Paris: Editions Les Monédièrers.

Mitchell, A. 2008. *Nazi Paris: The History of an Occupation, 1940–1944.* New York: Berghahn Books.

Mouré, K. and Schwartz, P. 2007. *On vit mal*: food shortages and popular culture in occupied Paris. *Food, Culture and Society*, 10, 261–95.

Mouré, K. and Grenard, F. 2008. Traitors, *trafiquants*, and the confiscation of illicit profits in France, 1944–1950. *The Historical Journal*, 51(4), 969–90.

Pétain, P. 1989. *Discours aux Français, 17 juin 1940 – 20 août 1944* [Speeches to the French People, 17 June 1940 to 20 August 1944], edited by J.-C. Barbas. Paris: Albin Michel.

Rist, C. 1983. *Une saison gâtée: journal de la guerre et de l'Occupation 1939–1945.* [A Spoilt Season: Journal of the War and the Occupation 1939–1945] Paris: Fayard.

Ruffin, R. 1979. *Journal d'un J3.* [Journal of a J3] Paris: Presses de la Cité.

Sanders, P. 2001. *Histoire du marché noir 1940–1946.* [History of the Black Market] Paris: Perrin.

Taylor, L. 2000. *Between Resistance and Collaboration: Popular Protest in Northern France, 1940–1945.* Basingstoke: Macmillan.

Thomas, E. 1995. *Pages de journal 1939–1944.* [Journal Pages] Paris: Viviane Hamy.

Vailland, R. 1948. *Playing for Keeps*, trans. by G. Hopkins. Boston, MA: Houghton-Mifflin.

Vallotton, G., and Vallotton, A. 1995. *C'était au jour le jour. Carnets (1939–1944).* [It was from Day to Day: Notebooks (1939–1944)] Paris: Payot.

Veillon, D. 1995. *Vivre et survivre en France 1939–1947.* [To Live and Survive in France 1939–1947] Paris: Payot.

Veillon, D., and Flonneau, J.-M.,eds 1996. *Le temps des restrictions (1939–1949).* [The time of Restrictions], special issue of *Les cahiers de l'Institut d'histoire du temps présent*, 32–3.

Chapter 14

Nutrition Education in Times of Food Shortages and Hunger: War and Occupation in the Netherlands, 1939–1945

Adel P. den Hartog

Introduction

Nutrition education, now common in most European countries, is generally directed towards the problem of obesity. However, it actually began in a time of scarcity when people had limited access to food. Nutrition education became part of the food policy in many European countries during the economic crisis of the 1930s and the Second World War.[1] For Nazi Germany, nutrition was of primary importance. From 1935 nutrition education was carried out by the *Frauenschaft* the official women's organization aided by its well equipped laboratory in Dresden.[2] In the Netherlands, nutrition education as a governmental responsibility began in 1941, during the occupation by Nazi Germany. It was initiated by senior Dutch officials of agriculture and health. A number of questions will be addressed here, including: why and by whom nutrition education was initiated; the nutritional message; the uneasy position of nutrition education in times of occupation; and the diminishing access to food and the widespread hunger in the winter of 1944–45. Consideration will also be given to the legacy of nutrition education and the early period of reconstruction after 1945. The main sources used are included in the footnote below.[3]

Food, Nutrition, Economic Crisis and the Threat of War

During the economic crisis of the 1930s, unemployment rose quickly from 100,000 in 1930 to 480,000 in 1936. The total household expenditure of the unemployed was 19 per cent less than that of the lowest income-earning households. The general nutritional situation of the unemployed could be summarized as nutritionally

---

1   Jensen 1994: 94-5, Hietala 1995: 160–61, Smith and Nicolson 1995: 308–10.

2   den Hartog 1947: 98.

3   Main sources used: the journals *Tijdschrift Voeding*, 1939–48, *Voeding en Hygiëne*, 1939–44, Nationaal Archief [National Archives], den Haag, 215.33 Inventaris van het Archief van de Gezondheidsraad, 1920–56.

unsatisfactory, and from a culinary point of view dull, but not a cause for alarm.[4] A group of leading domestic science teachers, specialists in cookery and nutrition, succeeded in convincing the Minister of Social Affairs to make funds available for educational activities with the women of unemployed households. The core of the message was how to manage a household with limited resources, with particular emphasis on the need for a healthy diet.[5] Nutrition education started as a private initiative and part of a home economics training programme, but with additional funding from the government. For the minister it served as a means to mitigate the harsh measures taken by cuts of social benefits. Because of urban and rural differences, two organizations were set up, one for the major cities and the other for the rural areas. The Foundation for Rural Household Education, known under its Dutch acronym HVP was very active and nutrition education became a core activity, with cookery courses being used as a vehicle to spread the nutritional message.[6]

The Occupation

In anticipation of a new, large-scale European conflict the Dutch government took measures for ensuring food security. In 1937 the State Bureau for Food Supply in Times of War, known under its acronym RBVVO, was created under the leadership of the agronomist Ir. S.L. Louwes (1889–1953). In the shadow of the coming war, the State Bureau for Food Supply made plans for implementing food self-sufficiency in order to be able to reshape agricultural production when needed. Food stocks were built up and the structure of a food rationing system developed. The main fear was being cut off from overseas food supplies, as had happened during the First World War.[7] The government, together with most of the population, believed that the country would not be drawn into the conflict, but remain neutral as it had in the earlier war. On 10 May 1940 the Netherlands were invaded by Germany and the country was occupied shortly after. The head of state, Queen Wilhelmina, and her government went into exile in England, and on 19 May 1940 Dr. Arthur Seyss-Inquart (1892–1946) was appointed *Reichskommissar* [Imperial Commissioner] the highest civil authority, of the occupied Netherlands. Germany's strategy was to leave the national and local administration intact and establish a regime of oppressive yet indirect rule in order to control the occupied nation and incorporate it into the German war economy.

The administrative unit *Ernährung und Landwirtschaft* [Nutrition and Agriculture] was created by the *Reichskommissar* to control agriculture and the supply of food. Throughout the occupation it was in daily contact with S.L.

---

4   den Hartog 1983: 33–8.

5   Oldenziel and Dorst 2001: 63–5.

6   Huishoudelijke Voorlichting ten Plattelande.

7   Trienekens 1985: 9–42. In Dutch: Rijksbureau voor de Voedselvoorziening in Oorlogstijd, RBVVO.

Louwes of the State Bureau for Food Supply. Louwes was an outstanding senior civil servant, who by his diplomatic skills and persistence managed to resist efforts to Nazify the food rationing system. The occupying authorities realized that a well-fed population was a precondition for civil order and high economic production and therefore in the interests of the Reich.[8] Soon after the Occupation the food rationing system was implemented and each person received a ration card together with coupons for buying food and non-food articles. By June 1940 bread, coffee and tea were already on ration and in the years to come nearly all other foods followed suit. Outside the food rationing system food could be obtained from the black market at high prices. For persons in hiding, such as persecuted Jews and political opponents of the Nazis, it was difficult to get food coupons. They relied on obliging food rationing officers who supplied clandestine coupons and the generosity of the people who were willing to share their meagre resources. However, this was not sufficient and once most basic foods were rationed, resistance groups were forced to raid distribution offices. [9]

The Bureau of Nutrition Education

Before the outbreak of the war discussions took place amongst senior officials of health and agriculture on the necessity of establishing a nutrition council. The occupation prevented the official creation of such an institution, but to avoid German intervention an informal nutrition advisory body was created within the National Health Council. The commission advised on nutrition surveillance, nutrient composition of food rations and the meals provided by soup kitchens. Soon the commission became known as the Nutrition Council. Under the influence of its leading member, the Senior Health Inspector Dr. C. Banning, the council was convinced that the population and in particular housewives needed information on how to make the best use of the food available.[10] For that purpose the Bureau of Nutrition Education was created in 1941 under the leadership of the physician Dr. Cornelis den Hartog (1905–1993).[11] From that point onward, nutrition education became a nationwide activity, alongside the already existing home economics education; it was directed at all housewives of every stratum of society. The bureau was staffed by domestic science teachers, who were specialists in cookery and nutrition. They were women with experience in home economics education. Indeed, apart from the director, the government accountant and messenger, the team was predominantly female, quite exceptional for a government institution at that time.

The main task of the bureau was to inform the Dutch population on how to eat a healthy diet by making the most of the very limited food resources. Nutrition

---

8   Trienekens 1985: 114-6, van Kamp 2005: 191–202.
9   Barnouw 1999: 13–15.
10 den Hartog 1989: 199–215.
11 Rigter 1991: 236–7.

education became a governmental responsibility and was highly gender specific: the main target group of nutrition education was, in the terminology of that time, housewives. Women were considered the gatekeepers of the household, responsible for food purchasing, preparation and the distribution of the meals to each household member. With the diminishing food availability this responsibility became a heavy burden on the shoulders of women; the proper feeding of their children and husbands was a real challenge. Skills were required in order to make the right choices, and the basic assumption was that knowledge and insight into healthy nutrition would lead to the required food behaviour. Additionally, these choices could be enhanced by practical information on appropriate food preparation techniques.[12] From the available sources one can conclude that both nutritionists and the nutrition teachers were aware that nutrition education in times of scarcity had to be approached tactfully with respect for the target groups. Many of the housewives came from different socio-economic back grounds, and the class differences between teachers and their audience could hinder getting the nutritional message across.[13]

The Nutritional Message and Channels of Education

Investigations into the role of vitamins in health dominated the research agenda in the years 1920–40. Therefore, it is not surprising that vitamins were a major component of the nutritional message disseminated by the newly established bureau. Because of food scarcity much attention was given to meat-replacement recipes (e.g. the use of beans) and later fuel-saving cooking techniques such as: one-dish meals; shortening the cooking time of vegetables, which saved vitamins as well; the soaking of beans before cooking; and use of the hay box. The main elements of nutrition education are summarized in Table 14.1. The channels of nutrition education were twofold: firstly, direct information was provided through nutrition folders, brochures and courses. Secondly, indirect information was disseminated to the population through training cadres of experts in social and health organizations and in provincial electricity companies. Since the 1930s electricity companies had organized cookery demonstrations for housewives to promote cooking on electric stoves, and in response to this, gas companies had reactivated their own earlier initiatives in this field.[14] In a number of major cities nutrition information shops were set up, which were easily accessible for housewives looking for practical information on food and nutrition.

12　*Leidraad voor voorlichtingscursussen.* Voorlichtingsbureau van den Voedingsraad, den Haag, 1941. Nationaal Archief, Inventaris 215.33, 465.

13　van't Hoog and van Overbeek, 1940–41.

14　Oldenziel and van Dorst 2001: 68–73.

Table 14.1   Main Elements of the Nutritional Message disseminated by the Bureau of Nutrition Education, 1940/41–46

- Protect vitamins with short cooking times; use vegetable water as an ingredient for soups and sauces
- Cook potatoes in their jackets to maintain the vitamin content and reduce wastage
- Be economical in food preparation: avoid kitchen waste
- Save fuel by preparing one pot meals and using the haybox
- Eat wholemeal bread
- Milk is indispensable for children and adults
- Vegetables and fruits are vital for a healthy diet, in particular uncooked vegetables
- Dried legumes are an important source of calories and proteins, and a useful meat substitute
- Flavour and variety are preconditions for a balanced nutrition

In 1941–2, two million folders and brochures with nutrition information and recipes were sold at nominal prices to the public. The bureau believed that free distribution would in the long run lead to a diminishing appreciation of their material. Each folder had an illustrated cover and efforts were made to make the text as readable as possible, so the draft folders were tested on housewives.

It appeared that the most effective way of getting the message across was through the combination of food demonstrations at household exhibitions and the sale of these folders.[15] The bureau collaborated with the two home economics education services and a number of nutrition folders were jointly published.

Nutrition press releases were sent weekly or biweekly to 1,300 organizations such as newspapers, weekly magazines and professional journals. The bureau estimated that each release reached about 350,000 households. Press releases had the same format as the folders: some information on nutrition often followed by a recipe. Draft releases were discussed with selected housewives and representatives from the agricultural sector, home economics education and health inspection. Local nutrition teachers assisted the bureau by adapting the press releases to the regional situation and then sending them to the local press. They advised the bureau on possible themes for future press releases.[16] The bureau also gave lectures and training courses at meetings of domestic science teachers. Particularly successful was the booklet *Guidelines for Present Day Nutrition* which sold 170,000 copies from 1941–46. The booklet was rather like a mini-nutrition manual directed at housewives with a strong emphasis on one pot meals, hotpots and main course

---

15   den Hartog and van Schaik 1948–49: 12.
16   Ibid.: 13.

Figure 14.1   Folder: Stressing the Need to Eat Winter Vegetables

Source: Bureau of Nutrition Education 1942.

soups because of the fuel scarcity. An examination of the content suggests that the booklet was probably mainly utilized by educated women.[17] In major cities nutrition information shops were set up in order to make direct contact with the target population and the bureau had its own nutrition stands at exhibitions frequented by housewives.

Table 14. 2 Channels of Nutrition Education of the Bureau of Nutrition Education 1941–46, Developed in close Collaboration with the Household Education Services*

Channel	Type of Activity
Direct nutrition education	
Folders, brochures	Some titles: *Preparing potatoes; Preparing salad and spinach; Tomatoes are healthy; Food deterioration; Food conservation; Skimmed milk; Winter vegetables; Sea food; Storing winter vegetables; Fuel saving; Home gardening; Garden herbs for balcony and garden; Utilization of leftovers; Wild food plants; Making emergency stoves; Preparing sugar beets; Preparing Tulip bulbs.*
Exhibitions	Stands in department stores; participation in household expositions; mobile exhibitions in smaller towns; education films.
Information shops	Amsterdam, Rotterdam, den Haag, Utrecht and IJmuiden (sea food).
Indirect nutrition education	
Nutrition education and training through cadres	Cooking lessons; training in health centres; rural associations' centres; electricity and gas companies.
Distribution of nutrition press releases	Major dailies and magazines

*   In 1943, 1.5 million folders were sold. The exhibitions were visited by 81,000 people and 24,341 attended the training courses.

Sources: den Hartog, 1946, Voeding en Hygiëne, 1943, den Hartog and van Schaik, 1965.

---

17   CBS 1947: 243. *Richtlijnen voor de hedendaagsche voeding*, 1941, prepared by Dr E.G. van 't Hoog and the well known cookery and nutrition teacher Martine Wittop Koning.

Cine films were made to support the information activities at meetings and home exhibitions, but the radio, which was fully controlled by the occupation forces, was avoided because the bureau did not want its nutritional message to be appropriated for propaganda purposes.[18] Apart from the bureau, a number of cookery books on how to eat well in times of war were published by private publishers; these were popular and written by well-known cookery and nutrition teachers.[19]

There were abortive attempts by a high-ranking official in the Dutch Nazi Party to Nazify the bureau. Ir Louwes and Dr. Banning succeeded in convincing the German authorities that such a change was not in the interests of the food rationing system. However a diplomatic gesture was required and in 1943 the director of the bureau was sent on a fact-finding visit to food institutions in Berlin.[20]

In the summer of 1940 the State Bureau of Food Supply decided to establish Central Kitchens (soup kitchens) for the provision of cheap and nutritious meals.[21] Gradually the Central Kitchens spread all over the country and proved to be crucial during the hunger winter of 1944–45.[22] The bureau was made responsible for the nutritional surveillance of the kitchens, including all hygiene matters. The meals they provided could only be obtained with the presentation of a food coupon.[23] In 1943 the bureau established an experimental kitchen in order to obtain a more practical insight into the preparation of tasty and nutritious meals at a household level based on the limited food available – this became operational in early 1944. Previously staff members of the bureau had experimented with the recipes at home in their own kitchens.[24] The experimental kitchen was indispensable during the hunger winter of 1944–45 when recipes had to be written for unusual foods such as sugar beets and tulip bulbs.[25]

Household Self Sufficiency, Unusual Foods and Nutrition Education

The Bureau of Nutrition Education and the two home economics education services advised households on how to grow edible plants in their gardens, and urban housewives without gardens were advised to cultivate garden herbs in pots on the balcony.[26] A new suggestion for many city dwellers was that of turning their flower garden into a vegetable plot. City councils made parks and urban wasteland

---

18   Montijn 1991: 126.
19   Ferro 2000, Grimm 2008.
20   den Hartog 1989: 140–41.
21   In Dutch, *Centrale keuken*. Trienekens 1985: 210–11.
22   Ibid.: 381
23   den Hartog 1941, 4551–2.
24   Verslag van den werkzaamheden van het Voorlichtingsbureau 1943. Nationaal Archief, Inventaris 215.33, 461.
25   van Schaik 1970: 540–43.
26   *Kweek zelf kruiden*, 1942, Nationaal Archief, Inventaris 215.33: 465.

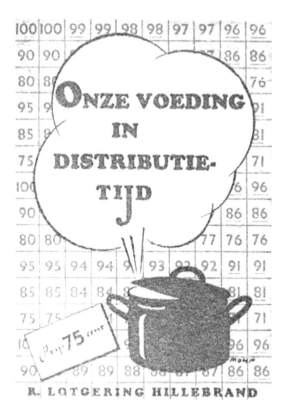

Figure 14.2 Cookery Book, 1941, with an Emphasis on One Pot Meals

Source: Ferro 2000.

available for food production and those with allotments could obtain seeds and fertilizers from the Ministry of Agriculture.[27] Trienekens estimated that the share of household self-sufficiency and related activities was no greater than 3 per cent of total food production, but for individual households it significantly assisted their survival despite the increasing problem of theft.[28] Similar measures were taken in other European countries as the food situation deteriorated.

Unusual food is always a relative concept, bound by chronological and cultural context. For example, boiling potatoes in their skins was very unusual in Dutch food culture at that time, but the folder issued on that topic stressed that peeling potatoes resulted in the loss of one fifth of their weight.

---

27 Barnouw 1999: 12–13, de Knecht-van Eekelen 2003: 127.
28 Trienekens 1985: 322–3 and 328.

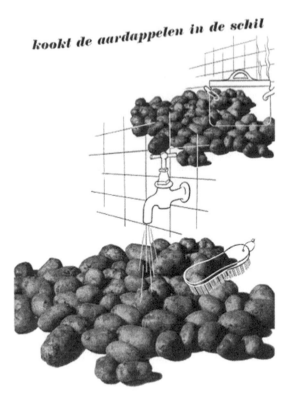

Figure 14.3   Folder: Boil Potatoes in their Jacket to Save Food and Vitamins

Source: Bureau of Nutrition Education 1942.

The folder tried to overcome the widespread aversion to the idea of eating potatoes and their skins: 'just try it' was the advice, 'the first day you may shudder, but on the seventh day even the master of the house will admit that it makes the potatoes very tasty'. Skimmed milk was unpopular and regarded as a drink for the poor, so a folder on the subject informed housewives that the value of skimmed milk should not be ignored. Despite the removal of the milk fats and with them the vitamins A and D, it remained an important source of proteins and calcium.[29] A press release advised the consumption of cow brains, as the bureau regretted the fact that people considered animal brains as useless offal when they contained

29   *Kookt de aardappelen in de schil*, 1942, *Taptemelk*, Nationaal Archief, Inventaris 215.33: 465.

plenty of fat and phosphorus and were also very tasty.[30] Many urbanites were not aware that numerous wild plants and berries found on roadsides, and in fields and woods were edible and of nutritional importance for their vitamin content, so a coloured brochure was distributed: *Our Wild Vegetables and Fruits.*[31] All these wild foods served as useful additions to the food ration.[32] For the same reason the Netherlands Mycological Association published a well-illustrated brochure on edible mushrooms that could be collected in the fields, although the brochure cautioned care in their selection given the dangerous consequences of errors.[33]

A popular activity in autumn was the search for beechnuts, which have a high fat content, in parks and woods. Unfortunately, in October 1942 some cases of poisoning were reported. While not deadly, poisoning had unpleasant symptoms including nausea, diarrhoea and dizziness. The bureau issued press releases with the advice to consume no more than 50 beechnuts, although the exact source of the toxins was never properly identified.[34] Scarcity followed by unavailability caused housewives and producers to seek out ersatz products: chicory was reintroduced to replace coffee beans and glucose syrup was used instead of sugar. However, the bureau did not generally concern itself with ersatz products, but concentrated its efforts on identifying alternative, less common foods for human consumption.

In the winter months of 1944–45, when the staple foods of wheat, rye and potatoes were vastly depleted in the western part of the country, the food authorities turned to the last resort of alternative supplies for human consumption: sugar beets, tulip and crocus bulbs. The bureau devised recipes in their experimental kitchen and passed their findings to the soup kitchen cooks and housewives. Other horticultural products were advised as being unsuitable for human consumption, for example the toxic hyacinth and daffodils bulbs. Regardless of the warnings issued they were still consumed, but the effect on those that ate them is not recorded. Bulbs of gladioli and iris were also eaten despite the fact that they tasted disgusting, and even dahlia bulbs were consumed, although even after an hour of cooking they remained hard and unpalatable.[35] Ironically, the tulip, whose flower was a symbol of Dutch warmth and cosiness, became a metaphor for food misery through the consumption of its bulbs. In other countries the hunger generated by warfare had compelled people to eat unusual food as well.[36]

---

30 Ook hersenen zijn een smakelijk voedsel, Voorlichtingsbureau, 3 April 1941, Nationaal Archief, Inventaris 215.33: 465.

31 *Onze wilde groente*, 1942.

32 van Schaik 1944–45: 7–8.

33 *Eetbare paddenstoelen*, 1941, Nationaal Archief, Inventaris 215.33: 468.

34 van Eekelen et al. 1943.

35 van Schaik 1944–45: 9–11.

36 See Chapters 7 and 9 by Maja Godina Golija and Pavel Vasilyev.

Coping with diminishing Food Supplies and Hunger

The rationing of potatoes, the staple food of the Netherlands, on 26 April 1941 made it clear to everyone that the war had had a profound impact on society. Nutrition education aimed to make the best of the situation and until September 1944 the food and nutritional situation was not a cause for alarm for most people, at least from a dietary energy point of view. Except for those in hiding, there was sufficient food to eat, but the daily meal became increasingly dull as the pleasures of the table were gradually stripped away.[37] From 1941 illegal food production and the black market became omnipresent, the latter gradually introduced inequality into the food rationing system. The more affluent classes were able to buy scarce products such as meat and dairy, but during the hunger winter 1944–45 these products were no longer available even to those with money.[38] In other countries such as France similar effects of the black market were observed.[39] The 1943 annual report of the bureau mentioned, without specifics, that the circumstances in which its activities were undertaken had been difficult; an indirect reference to the imposition of curfews, a shortage of transportation and a lack of basic teaching materials.[40]

In September 1944 the food situation in the western and most urbanized part of the country started to deteriorate rapidly. In June 1944 the Allied forces landed in Normandy, but in September 1944 Arnhem proved to be a bridge too far and above the Rhine the Netherlands remained occupied. In the meantime the Dutch government in London called for a national railway strike to frustrate military transports. In retaliation *Reichskommissar* Seyss-Inquart issued an order on 27 September placing an embargo on food transports to the major cities.[41] Nutrition as a weapon became a grim reality: there was no food and no fuel for cooking. An additional problem was the weather of October 1944 to March 1945, which was extremely bad: cold and wet, with sustained periods when the temperature dropped to minus 12C. In and around the cities all trees were felled, wood was taken from deserted houses to be used for simple one pit emergency wood-burning stoves which resulted in the mass reintroduction of one pot meals. The *Reichskommissar* revoked the embargo, but the transport of the provinces' food surplus to the cities by ships through the inland waters became impossible due to heavy frost.[42] The official food rations per head dropped from 1,800 kcal at the end of 1943 to 600 kcal during the winter of 1944–45.[43] The bureau struggled with the impossible task of informing the population on how to survive on a starvation diet. The

---

37 Trienekens 1995: 372–3.

38 Klemann 1997: 332–8.

39 See Chapter 13 by Kenneth Mouré.

40 *Verslag van den werkzaamheden van het Voorlichtingsbureau*, 1943.

41 de Jong 1991: 153.

42 Trienekens 1985: 382.

43 Burger, Drummond and Sanstead 1948: 5–6.

experimental kitchen became of crucial importance in the development of recipes for unusual foods, as has been discussed above.[44]

The society of the Western Netherlands was on the brink of total collapse: town-dwellers, mainly women, roamed the rural areas for food as men could be rounded up by the occupying forces and local Nazi militias for hard labour.[45] Food was bought by cash or barter, often at excessive prices. The food obtained was brought home by means of bicycles, carrier tricycles and handcarts, with the ever-present danger of being robbed.[46] Urged on by the Resistance, the London-based Dutch government asked for food aid from the Allies. This was a difficult request as the British government, and in particular Winston Churchill, feared the food could fall into German hands. After negotiations between Seyss-Inquart, the Dutch food authorities and secret contacts with London, in January 1945 the German authorities permitted the docking of two Swedish Red Cross ships loaded with flour and other food items at the northern port of Delfzijl. For ten days from 29 April, British and American planes dropped food onto specific sites in the Western Netherlands, but this was insufficient and on 2 May, road corridors for food transports were allowed through the German lines. Finally the occupation of the Western Netherlands ended on the 5 May 1945.[47] The estimated number of deaths in the winter of 1944–45 due to hunger and disease in the Western Netherlands is 16,000.[48]

Conclusion

What was the impact of nutrition education? It is not an easy question to answer. The number of nutrition folders and brochures distributed was impressive, estimated by the Central Bureau of Statistics at 10.3 million copies during the years 1941–45, to a total population of 9.2 million in 1945.[49] In essence the educational material was addressed to women, as housewives were responsible for the nutritional care of their household, in particular the children and the husbands. In view of these numbers the bureau must have reached large sections of the society, but little data is available to prove this. A CBS survey in 1944 on how households obtained information on food preparation could not be completed due to the Allied landings in Normandy. The scanty data showed that only 22.6 per cent of the respondents consulted recipes from newspapers and magazines, 23.9 per cent attended a training course in cooking and 31.1 per cent

---

44 Van Schaik 1970: 540–43.
45 Reuvekamp 1980: 14–20.
46 de Jong 1991: 244–66.
47 Barnouw 1999: 66–73.
48 Harts and Broekhuis 2007: 7.
49 CBS 1947: 234.

owned a cookery book.[50] However no questions were asked about the nutrition folders and brochures, and it is evident that the bureau made strenuous efforts to offer advice on nutrition.

After May 1945, food availability gradually returned to normal. By 1946, at a meeting of home economics education workers, the director of the Bureau of Nutrition Education stated that in contrast to recent experience, the problem of the future would be that of over-eating.[51] The experiences of the Second World War showed that nutrition education could make a valuable contribution to national health. During the second half of the twentieth century the Bureau of Nutrition Education remained an independent government funded agency as part of the state's food and health policy. Despite the many folders and brochures produced during the war, nutrition education lacked a clear nutritional tool or concept, so the bureau turned to other nations for inspiration. In the United States a very successful nutrition education tool was in use. It was called *Basic Seven*, and consisted of a disc explaining seven key food groups. Based on this example the bureau developed its own educational tool, adapted to Dutch food culture, which became the *Disc of Five* (1953). The disc was divided into five segments or food groups, representing all the required nutrient sources, and carried the message that by choosing food out of each group one could compose a healthy diet.[52] The concept behind the *Disc of Five* was that it was essential to be able to visualize the rather abstract nutritional message.[53] In the postwar years, learning from America in the field of nutrition became customary in other European countries as well such as in Germany.[54]

References

Barnouw, D. 1999. *De Hongerwinter.* [The Hunger Winter] Hilversum: Verloren.

Bruinenberg, S.N. and Winkelman, M.L.J. 1985. *Vijftig jaar voedingsvoorlichting HVP.* [50 Years of Nutrition Education HVP] Wageningen: Wageningen University Department of Human Nutrition.

Burger, G.C.E., Drummond, J.C. and Sandstead, H.R. eds 1948. Hunger and Starvation in the Western Netherlands: September 1944–July 1945. The Hague: General State Printing Office.

CBS. 1947. Economische en sociale kroniek der oorlogsjaren 1940–1945. [Economic and Social Chronicle of the War Years] Utrecht, De Haan.

---

50   van Schaik, 1946: 138–46.

51   Bruinenberg and Winkelman 1985: 64, 73.

52   den Hartog and van Schaik 1953: 251–5, Voorlichtingsbureau 1981: 2

53   In Dutch *Schijf van vijf.*

54   Thoms, 2004.

van Eekelen, M., den Hartog, C. and van der Laan, P.J. 1943. Vergiftiging door het eten van beukennootjes. [Beechnut intoxication] Nederlandsch Tijdschrift voor Geneeskunde, 87: 831–7.

Ferro, R.N. 2000. *Voeding in oorlogstijd.* [Nutrition in Wartime] Wageningen: Museum De Casteelse Poort.

Grimm, M. 2008. Het onderste uit de pan. Nederlandse kookboeken in oorlogstijd. [Dutchcookery books in times of war] *Geschiedenis Magazijn,* 43: 14–18.

den Hartog, A.P. 1983. Food habits in a situation of crisis: the unemployed and their food in the years 1930–1939 in the Netherlands. *Ernährungs-Umschau* Supplement, 30: 33–8.

den Hartog, C. 1946. De distributie in de jaren 1940 tot 1946. [Food rationing 1940–46] *Nederlandsch Tijdschrift voor Geneeskunde,* 90: 431–5.

den Hartog, C. 1947. Voedingsvoorlichting door de overheid in het buitenland. [Nutrition education in foreign countries] *Voeding* 8: 96–101.

den Hartog, C. 1989. *Daar komt de kippendokter.* [Memories] Arnhem: Repelsteeltje.

den Hartog, C. and van Schaik, T.F.S.M. 1946. Enkele gedachten over voedingsvoorlichting. [Some thoughts on nutrition education] *Voeding,* 7: 208–14.

den Hartog, C. and van Schaik, T.F.S.M. 1948–49. Beschouwing over de gebruikelijke methodiek bij de voedingsvoorlichting. [Views on the methodology used in nutrition education] *Voeding,* 9, 10, 200–204, 280–85, 23–30.

den Hartog, C. and van Schaik, T.F.S.M. 1953. Een nieuwe wijze van voedingsvoorlichting. [A new approach to nutrition education] *Voeding,* 14: 251–5.

den Hartog, C. and van Schaik, T.F.S.M. 1965. Vijfentwintig jaar voorlichtingsbureau voor de voeding. [25 years of Bureau of Nutrition Education] *Voeding,* 26: 398–419.

Harts, J.J. and Broekhuis, A. 2007. Reconstructie sterfte verloop tweede wereldoorlog. [A reconstruction of mortality figures of the Second World War] *Demos,* 23: 6–8.

Hietala, M. 1995. From famine to welfare: food patterns in Finland during the past hundred years, in ICREFH III.

van 't Hoog E.G. and van Overbeek, G.P.J. 1940/41. Practische voorlichting op voedingsgebied. [Practical nutrition education] *Voeding,* 2: 144–56.

Jensen T. O. 1994. The political history of Norwegian nutrition policy, in ICREFH II.

de Jong, L. 1991. Hongerwinter, [Hunger winter] in *Het koninkrijk der Nederlanden in de tweede wereld oorlog* [Kingdom of the Netherlands during the Second World War]. 's Gravenhage: Staatsdrukkerij, 10b, 153–266.

van Kamp, J.E. 2005. *Dien Hoetink. Biografie van een landbouw-juriste in crisis- en oorlogstijd.* [The Role of an Agricultural Laywer in Times of Crisis and War] Wageningen: Wageningen University unpublished thesis.

Klemann, H.A.M. 1997. De legale en illegale productie in de landbouw 1938–1948. [Legal and illegal agricultural production] *NEHA-Jaarboek*, 60: 307–38.

de Knecht-van Eekelen, A. 2003. From food to leisure: The changing position of the *jardin ouvriers* movement in the Netherlands from 1880 to the present day, in ICREFH VI.

Montijn, I. 1991. *Aan tafel: Vijftig jaar eten in Nederland* [50 years of eating in the Netherlands] Utrecht/Antwerpen: Kosmos.

Oldenziel, R. and van Dorst, C.J.M. 2001. De crisis: kapitaal- versus arbeidsintensieve techniek 1929–1940, [Crisis in the household 1929–1940] in *Techniek in Nederland in de twintigste eeuw* [Technology in the 20th century Netherlands], edited by J.W. Schot et al. Zutphen: Walburg Press.

Onze wilde groenten en vruchten. [Our wild vegetables and fruits] 1942. 's Gravenhage: Voorlichtingsbureau van den Voedingsraad.

Reuvekamp, A. 1980. Hongerwinter 1944–1945. Vier vrouwen halen herinneringen op aan haar hongertochten. [Hunger winter 1944–1945. Roaming for food: The memories of four women] *Opzij*, 8: 14–20.

Rigter, R.B.M. 1991. De gezondheids- en voedingsraad in oorlogstijd. [The Health and Nutrition Council during the war] *GeWiNa*, 14: 236–7.

van Schaik, T.F.S.M. 1944/45. Voedingsmiddelen door den mensch tijdens de oorlogsjaren 1940–1945 gebruikt, die voordien in Nederland vrijwel onbekend waren. [Unknown Foods consumed during the war] *Voeding* 6: 1–16.

van Schaik, T.F.S.M. 1946. Een enquête naar de voedselbereiding bij verschillende bevolkingsgroepen in augustus en september van het jaar 1944. [Survey on food prepration in 1944] *Voeding*, 7: 138–46.

van Schaik, T.F.S.M. 1970. Professor Dr. C. den Hartog en het Voorlichtingsbureau voor de Voeding. [Professor den Hartog and the Bureau of Nutrition Education] *Voeding*, 31: 540–43.

Smith, D. and Nicolson, M. 1995. Nutrition, education, ignorance and income: a twentieth- century debate, in *The science and culture of nutrition*, edited by H. Kamminga and A. Cunningham. Amsterdam: Rodopi.

Thoms, U. 2004. Learning from America? The travel of German nutritional scientists to the USA in the context of the Technical Assistance Program of the Mutual Security Agency and its consequences for West German nutritional policy. *Food History*, 2: 117–52.

Trienekens, G.M.T. 1985. Tussen ons volk en de honger: De voedselvoorziening, 1940–1945. [Food Supply in the Netherlands 1940–1945] Utrecht: Matrijs.

PART IV
War, Modernization and Innovation

Mikkel Hindhede and the Science and Rhetoric of Food Rationing in Denmark 1917–1918[1]

Svend Skafte Overgaard

'During the time of rationing, Denmark can, in other words, be considered as one big nature's sanatorium.'

Mikkel Hindhede, 1920.[2]

Introduction

Between 1840 and 1940 'food' was increasingly transformed into 'nutrition'.[3] Especially from 1900 onwards, the discourse of nutrition took off as part of a process of modernization that included an increasing rationalization of food, people and society. In this context, the First World War has been seen as a rupture and a catalyst for the growing influence of nutritional discourse which became part of popular culture and social regulation in Western Europe and the USA.[4] The construction of hunger[5] and the nutritional status of populations may be viewed as an integral part of 'biopolitics', in the sense of Michel Foucault.[6] As Corinna Treitel has argued in a German context, the biopolitics of 'rational nutrition' has yet to be researched in detail.[7] Taking Treitel's cue this chapter

---

1   This chapter is based on Svend Skafte Overgaard, Fra Mad til Ernæring. Mikkel Hindhede's ernæringsdiskurs og dansk madkultur 1900–1945. [From Food to Nutrition. Mikkel Hindhede's Nutritional Discourse and Danish Food Culture 1900–1945], University of Copenhagen Ph.D. 2008.

2   Hindhede 1920b: 58. [my translation]

3   Kamminga and Cunningham 1995: 1–14.

4   Levenstein 2003: 136–47, Smith and Phillips 2000, Teich 1995: 213–34, Vernon 2007.

5   Vernon 2007.

6   'Biopolitics' is used in the vein of Michel Foucault, focusing on the growing importance of the administration of processes of life as the raison d'être of politics and the modern state, Foucault 1980: 136–40, see Lemke 2009. For an analysis of the modern discourse of nutrition in Western culture, see Coveney 2006.

7   Treitel 2008: 1–2.

will explore the Danish rationing scheme during the First World War which was shaped by the advice of the controversial scientist and nutritional reformer Dr. Mikkel Hindhede (1862–1945). How did Hindhede come to influence the rationing scheme? And how did he make rhetorical use of the rationing period in the interwar period, staging the period from October 1917 to October 1918 as 'The Year of Health'?

Denmark 1914–18: Agriculture, Food and Rationing

When war broke out in 1914, Denmark declared its neutrality and it remained neutral for the entire duration of the war. As a result, Denmark managed to avoid most of the blockade initiatives implemented by belligerents until January 1917. Agricultural production suffered only a minor reduction and in terms of food production and provisioning, the situation was fairly stable until 1917.[8]

In January 1917, when Germany declared total warfare on all international shipping, foreign supplies of fertilizer and grains were cut off. This had a devastating effect on Danish agriculture. From 1914 to 1917 the number of pigs in Danish stables had been reduced by 15–20 per cent. Despite this reduction, the Danish life stock sector alone would need approximately 2,150 million kg of grains to maintain production in 1917 whereas Danish grain production for this year was estimated to be about 1,500 million kg to be shared among animals and humans.

The Danish government's response in the first months of 1917 included severe reduction of alcohol production, bread rationing and the cancellation of a regulation providing for the inclusion of barley in bread flour; the barley was to be reserved for pigs. Finally, the '*Kommissionen af* 4 April 1917' [Committee of 4th April 1917] was formed to deal with the distribution of the coming harvest of 1917. It consisted of experts, politicians and representatives from industry and agriculture and was charged to take both agricultural production and human nutrition into account.[9] This committee and its successor '*Ernæringsraadet*' [The Nutrition Committee], established in August 1917, came to hold the key position in the politics and administration of food during the remaining years of the war. Its expert advisors were professor of physiology at The Royal Danish Agricultural University, Holger Møllgaard (1885–1973) and Dr Mikkel Hindhede, head of Dr Hindhede's Research Institute for Human Nutrition from 1911 to 1932.

Principles of Nutrition in Denmark 1917–18

In a Commission report after the war Professor Møllgaard stated that the main question in 1917 had been to secure the 'sufficient' nutrition of the population

---

8 Christiansen 1989: 89–93, Helmer Pedersen 1988: 78.
9 Cohn 1928: 114–17, Kaarsted 1972: 140–67, Hindhede 1920b: 193. [my translation]

and that the main dilemma was to choose between lack of grain vs. lack of fats.[10] Looking at the numbers for grain production, there was seemingly no real dilemma, no real choice. A proportion of the animals had to be slaughtered in order to secure the food supply of the population. However, it took a certain gaze to see that – a gaze that accepted grain as a fully valid source of nutrition, which was not self-evident at the time.

Hindhede outlined the main principles guiding the food policy as follows:

1. A main principle of the planning was that any daily fare that offered sufficient nutrition and that was composed by natural foods always held the sufficient amount of protein. Vegetable protein held the same value as protein of animal origin.
2. We held that fat was a very important part of the daily fare (...), but not a physiological necessity (based on research showing vegetables being able to take the place of fats).
3. Bran was seen as a very valuable food that human beings were able to digest just as well as pigs (and almost as well as chewers of cud).
4. Supplementing the overall focus on a sufficient nutrition, we focused on a sufficient intake of vitamins. Thus, we prioritized milk, butter, wholegrain bread (including extra bran), grains of barley, potatoes, fruits and vegetables, but not meat, pork, pork fat, white bread or sugar. (When we still did give fairly big sugar rations it was only because sugar was a cheap and tasty food that was at the time easily obtainable).[11]

In this list Hindhede translated the main choice between fats and carbohydrates into a more explicit policy of nutrition. Hindhede's statement about the priorities of nutritional policy was directly modelled on the low protein, high carbohydrate fare which formed the core of Hindhede's own nutritional discourse.[12]

Mikkel Hindhede, Food and Health

Mikkel Hindhede has been characterized both as belonging to the anti-protein movement of the late nineteenth and early twentieth centuries, as a vegetarian and as an ardent rational nutritionist prioritizing economy over health.[13] He has also

---

10   *Beretning til Indenrigsministeren om Ernæringsraadets Virksomhed i Høstaaret 1917–18*, 1919: 7.

11   Hindhede 1920b: 40. [my translation] Hindhede also published on the subject in English, see Hindhede 1920a.

12   Overgaard 2008.

13   Baumgartner 1992: 74–5, 87, Carpenter 1994: 96–9, Lyngø 2003: 62, Merta 2000: 180–81, Meltzer 2003: 156.

been described as a 'puritan, joy-killing missionary'[14] who did not come to hold any lasting importance, as a fanatic agitator,[15] as well as an agrarian anti-modernist expressing a bodily engraved longing for authenticity typical of turn-of-the-century Danish modernity.[16] All these hold some validity. Hindhede was a multi-faceted figure in early twentieth century nutrition culture whose most important attribute was his ability to move between different discourses and institutional settings. This enabled Hindhede to position his version of nutritional truth favourably.[17]

Hindhede's discourse on food and health was initially formed through experience and self-experiments. It was articulated in a showdown with German nutritional authorities, foremost the protein centred discourse of Carl Voit (1831–1908), which Hindhede continued to project onto Voit's successors and above all Max Rubner (1854–1932). Hindhede also conducted long-term feeding experiments with human volunteers, which showed that human beings were able to live on as little as 25 g of vegetable protein per day. However, in practice he recommended 40–60 g daily protein consumption. According to Hindhede, the conventional modern diet ran counter to nature's call for simplicity. For Hindhede, this argument was confirmed by the excellent health of frugal small farmers in Western Jutland (Denmark), where he had grown up. In an effective rhetorical move, Hindhede turned the world on its head in terms of food and nutrition – and implicitly in terms of social and cultural hierarchies – by constructing a food pyramid that dismissed meat and protein in favour of calories, wholegrain, bread and potatoes.[18]

The Provision of Foods

On the advice of Møllgaard and Hindhede grains and potatoes were allocated from fodder to human consumption. This included 548 million kg of grain production, leaving only 1,000 million kg (half of normal requirements) for animal consumption. The number of pigs was reduced to 21 per cent of the pre-war figure (513,000) in February 1918 whereas only a smaller proportion of cattle were slaughtered.[19] In the wake of reduced nutritional quality of the fodder and a growing number of young animals, milk production fell from a prewar level of 3,500 million kg to 2,050 million kg per annum. Butter rationing was introduced and the weekly ration stood at 250 g per person. At the same time, 15 million kg of

---

14   Riismøller 1998: 133.
15   Mørch, 1987: 330.
16   Mellemgaard 2001: 290–303.
17   Overgaard 2008.
18   Hindhede 1907.
19   The number of pigs had already been reduced to about 80% of the pre-war-level in February 1917, Hindhede 1920b: 13.

butter were allocated for export, securing the provision of essential raw materials from abroad. The production of margarine was reduced to a minimum.

What did wartime restrictions mean in terms of actual food consumption? There was a rise in the consumption of wholegrain bread, which included 12 to 15 per cent extra bran normally fed to pigs, more porridge, more potatoes and less meat, pork, butter, margarine and white bread.[20] Importantly, alcohol consumption was reduced dramatically.[21] According to Hindhede, weekly consumption per person stood at 1,800 g coarse rye bread (including barley and bran), 700 g wheat bread, 219 g barley groats, 467 g sugar, 2,100 g whole milk, 700 g low fat milk, 175 g beef, 175 g pork, 250 g butter and margarine (almost exclusively butter), an unspecified number of eggs, 140 g cheese, 700 g low alcohol beer, 700 g of fruits and vegetables and 35 g coffee surrogate.

In terms of nutrition, this fare allowed for 2,300 kcals per person per day at the price of 0.70 kroner. The diet provided 68 g protein (27 g of animal origin) and 48 g fats. If these numbers were calculated to meet the norm of an average adult man the daily fare provided an intake of 2,900 kcals, with 82 g protein and 68 g fats. Of course there were variations across the population in terms of diet and nutritional intake. There were differences between social groups, between the sexes and between cities and countryside. Hindhede estimated that the energy intake in Copenhagen was lower than average, around 2,500 kcals. In the countryside the figure was higher than average, not least because of easier access to grains, potatoes, pork and milk and self-supplies, although this was difficult to calculate.[22]

Detailed discussions of the nutritional quality of the Danish diet in 1917–18 aside, there is no doubt that Denmark escaped hunger and malnutrition during rationing. Hindhede even declared the episode as an exceptional period of health in Danish and European history, naming the period October 1917 to October 1918 'The Year of Health'. He accredited rationing for the lowest mortality rate as yet, 10.4 per thousand, compared to 12.4 per thousand for the period 1910–1914 (according to Hindhede's numbers).[23]

The Danish Rationing Scheme and the Development of Food Policy in 1917–18

The most controversial part of food policy was undoubtedly the slaughter of pigs, which was carried out on Hindhede and Møllgaard's advice during 1917–18. Presented with the plan of The Committee of the 4 of April 1917, agricultural interests were split. The big pig farmers were faced with ruin when production was

---

20   Hindhede 1934: 118.
21   Hindhede 1920b: 21.
22   Ibid.: 17–20.
23   The technical details of statistics and connections between nutrition and health are not to be discussed here.

slashed and many years of breeding effort were being destroyed. Many were able to see the need for change but Thomas Madsen-Mygdal (1876–1943), who served as prime minister for the party '*Venstre*' (the long time liberal opposition party based in agriculture) from 1926–29, argued that it would be wise not to proceed with the dramatic reduction proposed. Smallholders were more supportive of Hindhede and Møllgaard's plan, and in the end, the pig framers were forced to give in.[24]

Hindhede had succeeded in getting funding for conducting nutrition research in his own laboratory from 1911, gaining a stamp of approval from the authorities. Nonetheless Hindhede continued to be a highly controversial figure, not least among his medical colleagues. According to *Ugeskrift for Læger* [Doctors Weekly], Hindhede's ideas were, 'Fanatical and against all previous experience'. They were not fit to 'convey the basis for the regulation of the nutrition of the people at a time when only the well proven and varied experience ought to lead the way.[25] How did Hindhede come to dominate food policy in 1917–18 despite his controversial position among doctors and the public?

The answer to this question can be found in multiple arenas which together form a matrix that put Hindhede in a privileged position in terms of the production and distribution of nutritional knowledge. Hindhede's inclusion in the '*Kommiteen af 4. April 1917*' has to be understood against the background of earlier developments in international scientific discourse on nutrition, the social and cultural turmoil and reform politics in Denmark and, finally, the peculiar circumstances of the First World War which were a perfect fit for Hindhede's ideas.

An International Perspective: Influence of Scientific Advisors on Wartime Food Policy in Germany and Britain

As Mikulas Teich, James Vernon and Corinna Treitel have pointed out, the influence of scientists as expert advisors in planning and executing food policies during the First World War in Germany and Britain was surprisingly modest in view of the standing of nutritional science and rational nutrition before 1914.[26] Teich argues that a higher social standing of scientists in Britain resulted in a larger influence of their views in the implementation of a nutrition policy than in Germany where the media, the army and economic interests of different groups influenced the cause of events markedly.[27]

In Britain competing discourses on how to feed the population most rationally complicated the issue and micronutrients (vitamins etc.) received a higher priority than in Germany where food value as measured in calories set the standard not leaving much room for considerations of vitamin content. Within the wider

---

24   Helmer Pedersen 1988: 78.
25   'Nytaar 1918', *Ugeskrift for Læger*, 1, 1918, 2–3.
26   Teich 1995, Treitel 2008: 21–3, Vernon 2007: 91–6.
27   Teich 1995: 223–7.

framework of a quantitative, thermodynamic calorific discourse, the norm of a healthy diet in Germany was influenced by Voit's long standing nutritional standard stressing the importance of protein; recommended intake stood at 118 g per man per day. In Britain the corresponding protein norm was set at 105 g.[28] Differences aside, the point here is that despite the rising importance of nutrition science, it was not self evident that food policy followed the advice of scientific experts during the First World War. Further, different standards promoted different arguments and responses with regard to food policy, planning and administration.

Compared to Britain and Germany, several factors have to be considered to explain the Danish situation. Denmark was a neutral country and thus not directly involved in the war. Therefore, there were no generals demanding meat for their troops. Denmark was a small country with a comparatively simple administrative structure and straightforward regional and national interests. A certain degree of political unity across the party divide characterized Danish politics during the First World War. The dominant economic interests of farming, production and trade were committed to this unity due to corporate devices and ideologies.[29] Furthermore, the scientific community in Denmark was small, reducing the complexity of competing views, sites of debate and strongholds of power. Finally, Denmark did not enter a crisis of resources until 1917 when there was much to be learned from the conditions elsewhere. Still, how did Hindhede come to hold the position as advisor to the government in 1917–18?

Mikkel Hindhede's Influence on Food Rationing: Socio-cultural and Discursive Framing

When Hindhede entered public debate on nutrition in 1906 he immediately attracted attention from the media. His books sold in large numbers in Denmark and were soon translated into several languages. On the one hand, Hindhede was firmly grounded in science and had an attentive public among domestic scientists and others interested in questions of food, nutrition and social reform. On the other, Hindhede's semi-vegetarian discourse was typical of early twentieth century holistic reform movements. He was portrayed as one of many health gurus of the time and much ridiculed in satire and elsewhere.[30]

Hindhede's discourse on food and health must be seen in conjunction with the profound social, cultural and political turmoil of the period which heightened

---

28   Teich 1995: 223–7. Vitamin research was only in its infancy in the period. However, competing discourses in Britain resulted in a complex discussion and less clear cut, authoritative advice from scientific experts to politicians than in Germany. Nevertheless, vegetables and vitamins were held in higher regard compared with Germany, see Teich 1995.

29   Christiansen 1989: 89–93.

30   Overgaard 2008.

the meaning of food, the body and health as sites of identity.[31] In order to understand Hindhede, his ideas on food and nutrition, their reflection of society and their impact in Danish food culture, the clash between city and countryside in Hindhede's discourse is an important gateway.

Not only did a frugal life in the countryside serve as Hindhede's ideological utopia, his nutritional discourse was in some respects directly interlinked with agro science. In 1905–8 Hindhede was a prominent participant in a fierce debate on the feeding of cattle. He defended fodder trials, guided by the principles of a low protein turnip diet, which had been developed by his uncle, the legendary veterinary, researcher and inventor of the modern centrifugal separator for milk N.J. Fjord (1825–91).[32] An article in the social liberal Copenhagen newspaper *Politiken* expressed the spirit of Hindhede's adversaries who included the later Nobel Laureate, zoologist and physiologist, August Krogh (1874–1949), 'The conflict over the milking cow has become symbolic and mythological. It is the conflict between light and darkness in our public life, between the present and the future we are building in Denmark'.[33]

In Hindhede's framing of events, the conflict between past and future, light and darkness was turned on its head in this interpretation of things. The cultural, political, economic and bodily hierarchies of the nineteenth century, which Hindhede saw expressed in the critique of the fodder trials, were to be replaced by a new healthy order based on a lifestyle modelled on the hard-working peasants and farmers of western Jutland. For Hindhede, self-reform and food reform were the core of a new order of society. Scientific investigations with grounding in practical life were to marry self-discipline, resulting in happiness, the collapse of old hierarchies, false pretensions and many of the ailments that characterized life in modern society.

Hindhede's experiments and his participation in the debate over the feeding of the dairy cow resulted in financial backing from agricultural associations when Hindhede needed money to fund a new research institute for human nutrition, which functioned from 1911–32. He succeeded in obtaining this funding by appealing to the minister of justice, who was part of the same agro-cultural social and ideological groups that formed Hindhede's hinterland. In part, his efforts to obtain funding were successful because he used the language of leading administrators and reform politicians. Hindhede transferred the rational economic calculus of input and output taken from the cattle experiments to the area of human nutrition

---

31   Hau 2003: 1–3, Kerbs and Reulecke 1998: 10–18, Krabbe 1998: 73–5, Meyer-Renschausen and Wirz 1999: 323–41, Stolare 2003, Whorton 1982.

32   The debate dominated the pages of popular agricultural magazines as *Landsbladet*, but was also conducted in leading Copenhagen newspapers *Politiken* and *Berlingske Tidende*.

33   'Personlige Formaal', [Personal aims] *Politiken* 27 September 1905.

transferring N.J. Fjords's agro-scientific concept of 'feeding units' to the concept 'food units' using the ratio between calories and price as a decisive norm. [34]

Hindhede was not alone in transforming the value of foods into a rationally calculable ratio of nutritional value vs. economy.[35] However, his application of the language of rational nutrition and his invocation of an agricultural setting effectively enlarged the message in a context where the distribution of social, political and cultural power was drifting from an urban bourgeoisie, an academic, economic, social and political elite to a rising, socially mobile, group based in the values of agriculture, democracy, popular education, 'practical science' and social reform.

Hindhede's discourse of rational nutrition resonated with Ove Rode, the powerful Minister of the Interior who is credited as the main instigator of the modern welfare state in Denmark. Defending Hindhede in parliament, Rode thus reified the discourse of food as nutritional resource:

> The fact that the grain based foods such as bread and barley porridge have held a prominent position during these hard times has been of the utmost importance.. if we had been sloppy with our sources of nutrition, catering to taste and habit, well, then we would have led our country into great trouble.[36]

According to Rode, Denmark conquered the challenges of food shortages by putting the head before the stomach.[37] Food was translated into 'nutrients' using a framework that conceptualised nutritional resources in terms of calories and that dispensed with the need for meat that used to be grounded in science and food culture alike. However, although a politics of crisis, Rode's acceptance and active use of the implications of the discursive framework of Hindhede's nutritional science points towards a re-conceptualisation of food politics in times of peace as well. As Rode forecast the coming of a bigger state and the regulation of society in general, he also forecast the coming of a peacetime food policy based in nutrition science and social economy.

Hindhede had been much debated in the Danish parliament from 1909 onwards. Several voices continued to characterize Hindhede as a fanatic and to oppose the frugality of his ideas. However, whereas the mere idea of conceptualizing 'food' as nutrients and intervening in the food ways of the population was scoffed at in 1909–13, this was increasingly accepted after 1918 and throughout the interwar years.[38]

---

34   Hindhede 1906: 115.

35   Treitel 2008.

36   *Rigsdagstidende. Forhandlinger paa Folketinget 1918–19*, c. 367. [my translation]

37   The intricacies of the body-mind split in relation to food and nutrition is discussed in Coveney 2006.

38   *Rigsdagstidende. Forhandlinger paa Folketinget 1923–24*, c. 3192–237.

How far the conceptualization of food and health had moved was expressed in a 1935 parliamentary debate. Hindhede was now criticized for paying too little attention to vitamins. Even among Hindhede's supporters this was acknowledged. As Minister of the Interior Berthel Dahlgaard put it in defence of Hindhede, 'One has to remember that Hindhede worked in – we may call it – the age of calories. Now we are living in the age of vitamins'.[39] The debate over vitamins and the general acceptance of the concept of nutrition in parliament that emerged during the interwar period coincided with a wider nutrition discourse in society. Therefore, the thesis that the First World War can be seen as a turning point and a forerunner for an intensification of nutrition discourse in general and government food policy initiatives in particular also seems to make sense in the case of Denmark.[40]

'The Year of Health' and Interwar Nutrition Culture in Denmark

Hindhede named the period from October 1917 to October 1918 'The Year of Health' producing a number of statistics showing how the Danish low protein, low alcohol and high fibre diet had resulted in the lowest mortality rate as yet (10.4 per thousand).[41] Combining his lessons learned from feeding experiments with cattle, small-scale but prolonged clinical feeding trials with human volunteers and the collection of food data from households, he staged the health of the Danish population 1917–18 as a large scale test case of his nutritional theory. Simultaneously gloating over what he characterized as the fallacies of German nutrition science and administrative efforts during the war,[42] Hindhede framed the question of nutrition as part of a dichotomy between a natural and an unnatural way of life. This allowed him to question the modern lifestyle and the modern diet;

> If the Danish population in general consisted of healthy people living a normal life, the remarkable results of the rationing period would be impossible to understand. However, if we consider modern people as predominantly overfed and poisoned patients, then the change is somewhat easier to understand. For a year, we have put all these people on a diet, for which they did not quite volunteer.

39　*Rigsdagstidende. Forhandlinger paa Folketinget 1935–36*, c. 1797. [my translation]

40　Levenstein 2003: 136–47, Teich 1995, Treitel 2008: 22–3, Vernon 2007: 94–6.

41　Hindhede chose this period, to avoid the impact of the Spanish flu that started to affect mortality in Denmark from October 1918. He also allowed time for the changed diet, introduced earlier in 1917, to take effect. Of course this can be, and was, criticized for a number of reasons, but the technical intricacies of the statistics involved are not to be considered here. Nor is the question of mortality rate as an indicator of nutritional health or the relation between the short time span between the beginning of the nutritional changes and the changes in health of the population to be discussed in detail.

42　Hindhede1916: 445–9, 471–5, 501–5, 534–41, Hindhede 1920b: 17.

> During the time of rationing, Denmark can, in other words be considered as one big nature's sanatorium.[43]

Hindhede's agitation for 'The Year of Health' continued to stir debate. In the scientific community, among others, Max Rubner was furious.[44] Hindhede's colleagues among Danish doctors questioned his calculations. As vitamins acquired increasing importance in the interwar years, Hindhede was criticized for underestimating the contribution of meat, butter and whole milk as providers of vitamins.

In the context of a growing attention to questions of health and nutrition by the Danish public, 'The Year of Health' was translated into 'The Year of Butter' by Dr Johanne Christiansen, Hindhede's most ardent adversary in Denmark in the interwar years. According to Christiansen, there might have been an improvement in health during the period October 1917 to October 1918, but this was due to reduced alcohol consumption and the substitution of butter for margarine. Christiansen thus argued in the vein of the 'newer nutrition' of the interwar years, focusing on the vitamin A and D content of the diet.

Christiansen's critique of Hindhede formed part of a debate in the 1930s about the nutritional condition of the Danish population, not least the country's youth, which was said to be crippled by a frugal diet, low in meat, butter and whole milk.[45] Another issue was a 'margarine war' between Hindhede and Christiansen who debated the pros and cons of butter and margarine. This finally resulted in the fortification of margarine with vitamins in Denmark from December 1936.[46] Christiansen invoked the strong physiology of hunters and gatherers and a mythological northern past prior to the introduction of agriculture, hence valuing meat, blood, butter and whole milk.[47] Hindhede, on the other hand invoked a more recent past: the simple farmer's life before modernization who lived on potatoes and whole grain bread. He claimed that vegetable foods and skimmed milk were sufficient vitamin providers.[48] Both continued to invoke the rationing episode of 1917–18.[49]

Moving with ease between different arenas of meaning in the construction of popular truths on food and health, the debates between Hindhede and Christiansen can be seen as exemplifying the widening scope of nutritional discourse in the interwar period. This is also the case when compared to the more moderate voices of the time, who placed Hindhede and Christiansen at the outer poles of nutritional discourse in Denmark. This discourse pointed at the complexity of the relationship

---

43   Hindhede 1920b: 58.

44   Rubner 1930.

45   Rømer 1989.

46   Among others, the nutrition- and margarine war was waged in *Politiken* 15, 27, 28 and 30 January, 2 February, 7, 15, 21, 24, 27 March, 12 November 1936.

47   Christiansen 1935.

48   Hindhede 1934.

49   Hindhede1938 (an answer to Christiansen's recent critique).

between food and health and the importance of negotiated consent in the area of human nutrition, which found expression in new national and international committees of nutrition.

Conclusion

The First World war marks a disruptive moment in the history of modern nutrition culture. For the first time, the planning and administration of the diet of the entire population was framed by the language and mode of thought of nutrition science.[50] In Denmark the rationing scheme of 1917–18, was largely due to the advice of controversial scientist and food reformer Dr Mikkel Hindhede. Eighty per cent of pigs were slaughtered and grains were transferred to human nutrition. Numbers of dairy cattle were only reduced to a minor degree, allowing for a continued substantial production of milk and butter. Hence, the Danish diet was based largely on grains, potatoes, vegetables, milk and butter. Denmark came out of the war without hunger and malnutrition.

This chapter has examined how Hindhede managed to get into a position where he came to hold the privileged position as advisor to the government. Hindhede acquired this position for a number of reasons grounded in the socio-cultural circumstances of Danish politics and in Hindhede's ability to speak the language of power in a very broad sense. Hindhede always portrayed himself as an outsider, and his contemporaries partly did the same. However, this position as an outsider was more or less framed as a conflict between conservative forces of bourgeois society and progressive values of forward thinking farmers, practical scientists and reform politicians. Hindhede was part of an up-and-coming social and cultural milieu that was gaining hold in the first decades of the twentieth century. In the sense of Pierre Bourdieu, his habitus[51] played an important part in his ability to speak a language that appealed to the right people at the right time, positioning Hindhede as a privileged agent of nutritional discourse.[52]

Pointing at the lowest mortality as yet in Denmark, Hindhede staged the period October 1917 to October 1918 as 'The Year of Health'. Denmark was presented as a large scale nutritional experiment that proved his theories of a high fibre, low protein diet to be correct. Subsequently and particularly in the 1930s, Hindhede's 'Year of Health' was criticized. This was due to the rising importance attached to vitamins and in a new context where his frugal ideas of abstention were increasingly surpassed by an appreciation of a rich and varied diet, including expensive meat, butter and whole milk. Both in its form and in its content, the debate is suggestive of how nutrition discourse was changing and widening its scope to include a broad range of settings, from intensified administrative efforts and regulations in the

---

50   Teich 1995, Treitel 2008, Vernon 2007.
51   Bourdieu 1990: 9–18.
52   Overgaard 2008.

spirit of national nutritional politics and international committees to an enlarged space for public debate that encompassed concepts of nutrition science as well as ideological projections of the past and the present on the eating body.

References

Baumgartner, J. 1992. *Ernährungsreform – Antwort auf Industrialisierung und Ernährungswandel.* [Reform of Nutrition – A Response to Industrialisation and Dietary Change] Frankfurt am Main: Peter Lang.

Bourdieu, P. 1990. *In other Words.* Stanford: Stanford University Press.

Carpenter, K. J. 1994. *Protein and Energy. A Study of Changing Ideas in Nutrition.* Cambridge, New York and Melbourne: Cambridge University Press.

Christiansen, J. 1935. *Mad er Mandens Kraft.* [Food is the Power of Man] København: Gyldendalske Boghandel – Nordisk Forlag.

Christiansen, N. F et al. 1989. *Danmarks Historie.* bd. 7 [History of Denmark vol. 7] København: Gyldendal.

Cohn, E. 1928. *Danmark under den Store Krig.* [Denmark during the Great War] København: G.E.C. Gads Forlag.

Coveney, J. 2006. *Food, Morals and Meaning.* London: Routledge.

Foucault, M. 1980. *The History of Sexuality: An Introduction.* New York: Penguin.

Hau, M. 2003. *The Cult of Health and Beauty in Germany: A Social History 1880–1930.* Chicago: University of Chicago Press.

Helmer Pedersen, E. 1988. *Det danske landbrugs historie 4: 1914–1988.* [History of Danish Agriculture vol. 4] København: Landbohistorisk Selskab.

Hindhede, M. 1906. *En Reform af Vor Ernæring.* [A Nutritional Reform] København: Gyldendalske Boghandel.

Hindhede, M. 1907. *Økonomisk Kogebog: Praktisk Ernæring.* [Economical Cookbook: Nutrition in Practice] København: Gyldendalske Boghandel.

Hindhede, M. 1913. *Protein and Nutrition: An Investigation.* London: Ewart, Seymour & Co.

Hindhede, M. 1916. Die Ernährungsfrage. [The question of nutrition] *Berliner Klinische Wochenschrift,* 53, 445–9, 471–5, 501–5, 534–41.

Hindhede, M. 1920a. The effect of food restriction during the war on mortality in Copenhagen. *Journal of the American Medical Association,* 74 (6), 381–2.

Hindhede, M. 1920b. Beretning fra M. Hindhede's Kontor for Ernæringsundersøgelser. Beretning til Indenrigsministeriet om Rationeringens indvirkning paa Sundhedstilstanden 13 [The Effect of Rationing on Health. Report to the Minister of the Interior 13] København: Jacob Lund. Medicinsk Boghandel.

Hindhede, M. 1934. *Fuldkommen Sundhed og Vejen Dertil.* [The Road to Perfect Health] København: Gyldendal.

Hindhede, M. 1938. Nutrition in Denmark during the war. *British Medical Journal*, 1, 1339–40.

Kaarsted, T. 1972. *Indenrigsminister Ove Rodes dagbøger 1914–18*. [The Diaries of Ove Rode, Minister of the Interior] Oslo: Universitetsforlaget.

Kamminga, H. and Cunningham, A. 1995. Introduction: The science and culture of nutrition, 1840–1940, in *The Science and Culture of Nutrition, 1840–1940*, edited by H. Kamminga and A. Cunningham. Amsterdam: Rodopi, 1–14.

Kerbs, D. and Reulecke, J. 1998. Einleitung, in *Handbuch der deutschen Reformbewegungen 1880–1933* [Introduction, Handbook of the German Reform Movement], edited by D. Kerbs and J. Reulecke. Wuppertal: Peter Hammer Verlag, 10–18.

Krabbe, W.R. 1998. Lebensreform/Selbstreform, in *Handbuch der deutschen Reformbewegungen 1880–1933* [Life Reform/ Self Reform, Handbook of the German Reform Movement 1880–1933], edited by D. Kerbs and J. Reulecke. Wuppertal: Peter Hammer Verlag, 73–5.

Lemke, T. 2009. *Biopolitik – en introduktion*. [Biopolitics – An Introduction] Copenhagen: Hans Reitzels Forlag.

Levenstein, H. 2003. *Revolution at the Table*. Berkeley and Los Angeles: University of California Press

Lyngø, I. J. 2003. *Vitaminer – Kultur og vitenskap i mellomkrigstidens kostholdspropaganda*. [Vitamins – Culture and Science in Interwar Propaganda of Nutrition] Oslo: Universitetet i Oslo.

Mellemgaard, S. 2001. *Kroppens Natur: Sundhedsoplysning og naturidealer i 250 år*. [The Nature of the Body: Health Education and Ideals of Nature for 250 Years] København: Museum Tusculanums Forlag.

Meltzer, J. 2003. *Vollwerternährung: Diätetik, Naturheilkunde, National-sozialismus, sozialer Anspruch*. [Whole Foods: Dietetics, Natural Healing, National Socialism, Social Demands] Stuttgart: Frantz Steiner Verlag.

Merta, S. 2000. Keep fit and slim! Alternative ways of nutrition as aspects of the German health movement, 1880–1930, in ICREFH V, 170–202.

Meyer-Renschausen, E. and Wirz, A. 1999. Dietetics, health reform and social order: Vegetarianism as a moral physiology: the example of Maximilian Bircher-Benner (1867–1939). *Medical History*, 43, 323–41.

Mørch, S. 1987. *Den Ny Danmarkshistorie*. [The new History of Denmark] København: Gyldendal.

Overgaard, S. S. 2008. Fra Mad til Ernæring: Mikkel Hindhede's ernæringsdiskurs og dansk madkultur 1900–1945. [From Food to Nutrition: Mikkel Hindhede's Nutritional Discourse and Danish Food Culture 1900–1945] Ph.D. Thesis, University of Copenhagen.

Riismøller, P. 1998. *Sultegrænsen*. [The Hunger Line] København: Nyt Nordisk Forlag.

Rubner, M. 1930. Über den Niedrichsten Eiweissverbrauch. [On the limit of protein requirements] *Clinical and Experimental Medicine*, 72(1), 99–118.

Rømer, H. 1989. Bøffen og Bananen – Hindhede-Christiansen-debatten i mellemkrigstiden [The steak and the banana – The debate between Hindhede and Christiansen in interwar Denmark]. *Den Jyske Historiker*, 48, 89–106.

Smith, D. F. and Phillips, J. eds 2000. *Food, Science, Policy and Regulation in the Twentieth Century: International and Comparative Perspectives.* London: Routledge.

Stolare, M. 2003. *Moderniseringskritiska rörelser i Sverige 1900–1920.* [Anti-Modern Movements in Sweden 1900–1920] Karlstad: Avhandlingar från historiska institutionen i Göteborg.

Teich, M. 1995. Science and food during the Great War: Britain and Germany, in *The Science and Culture of Nutrition, 1840–1940*, edited by H. Kamminga and A. Cunningham. Amsterdam: Rodopi, 213–34.

Treitel, C. 2008. Max Rubner and the biopolitics of rational nutrition. *Central European History*, 41, 1–25.

Vernon, J. 2007. *Hunger: A Modern History.* Cambridge, MA: Harvard University Press.

Whorton, J. C. 1982. *Crusaders for Fitness: The History of American Health Reformers.* Princeton, NJ: Princeton University Press.

Chapter 16

The Modernization of the Icelandic Diet and the Impact of War, 1914–1945

Guðmundur Jónsson and Örn D. Jónsson

Introduction

As the nineteenth century progressed the process of globalization, bringing freer movement of goods and integration of markets, had tremendous impact on European food cultures. In few countries was this process as discernible as in Iceland, a poor and remote country in the North Atlantic. As Iceland's foreign trade gradually expanded and its economy opened up, its food culture was transformed in a space of only a century from the end of the Napoleonic wars to the outbreak of the First World War.

The chief characteristic of this transformation was a steady shift from an animal-based to a grain-based diet, a shift that was opposite to the general trend in Europe.[1] As modernization of the diet was closely linked to stronger ties with the outside world, especially imports of cheap carbohydrates for the growing urban population, it was heavily dependent on free and uninterrupted foreign trade. This chapter examines the shift from a traditional to a modern food culture and how the two world wars contributed to this process. We seek to explain the dietary changes in Iceland with reference to the economic dynamics of food production and consumption as well as the country's political circumstances. Iceland was part of the Danish realm until 1944 and it remained neutral during the two world wars. We pay special attention to the emergence of a new food culture associated with urbanization and the rise of the market economy, and the tensions between the nascent urban population and the farming community over various food policy issues.

The Icelandic 'Carbohydrate Revolution'

Icelandic food history has some peculiar features. For centuries, about 90 per cent of food consumed was of animal origin whereas cereals were of little importance.[2] Milk, butter, mutton, suet, fish and other animal-based foods dominated the Icelandic diet to an extent almost without parallel in Europe except perhaps

---

1 Grigg 1995.
2 G. Jónsson 1998.

among the nomads in far-northern Europe, e.g. the Sami in Lapland and the Inuit in Greenland. In the second half of the nineteenth century when most of Europe was turning away from 'starchy staples' in favour of more animal-based foods, Iceland was moving in the opposite direction.[3]

Until the twentieth century, struggling for survival was a way of life for most Icelanders.[4] The country was one of the poorest in Western Europe and the great majority of its population subsisted on animal husbandry and fishing. Foodstuffs and utensils that characterized continental food traditions and consumption patterns were often absent. Since Icelanders did not grow grain, all cereals had to be imported and were relatively expensive compared to domestic foodstuffs. Barley had been grown in the Middle Ages but by the sixteenth century the practice had died out as a result of a colder climate. Another striking feature of the Icelandic diet was the shortage of salt. Icelanders had to resort to other food preservation methods, most importantly pickling in a saltless brine of fermented whey. Meat, blood puddings and butter were stored in whey, which preserved the products and kept them edible over the winter months. A scarcity of fuel also shaped the diet in many ways as it made baking difficult or nearly impossible.[5] Bread was therefore a luxury and cereals were eaten in the form of gruel. Most other foods were eaten cold, the dried fish (*stockfish*) with sour butter, blood puddings, milk and the all-important *skyr* – a special Icelandic type of curd. Families did not eat their meals at tables except on special occasions, instead each member sat on his or her bed in the common sitting-sleeping room while eating cold food from an *askur*, an all-purpose wooden dish carved out of piece of wood with a lid. Meat was scarce and vegetables were curiously absent until the end of the nineteenth century. Sporadic attempts to grow potatoes had been made since the seventeenth century, but their significant cultivation and consumption did not start until the end of the nineteenth century.[6]

An important change began in the second quarter of the nineteenth century when cereals, rye and barley in particular, were becoming cheaper. Imports soared and gradually led to a shift in consumption away from an animal-based towards a more grain-based diet. The 'carbohydrate revolution' was underway in Iceland. It accelerated towards the end of the century with a rapid increase in sugar imports and the belated popularity of the potato. Figure 16.1 shows this shift over the course of more than a century and a half. Imported foodstuffs, which were overwhelmingly of vegetable origin, became more important in the diet while domestic foodstuffs, mostly of animal origin, became relatively less important.

The driving forces of this dietary transformation were improved transportation and changes in relative prices within the framework of traditional agricultural society. Growing foreign trade strengthened the ties with Europe, in particular Denmark, Iceland's traditional gateway to the wider world. Danish food culture

---

3   Grigg 1995.
4   Ö.D. Jónsson 1994.
5   Gísladóttir 1994. See also Gísladóttir 1999: 65–73.
6   Gísladóttir 1999: 299–303, G. Jónsson 1998: 35–6.

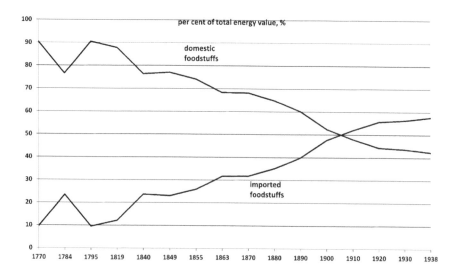

Figure 16.1   Food Supply in Iceland: Domestic Production vs. Imports, 1770–1938 (per cent of total energy value, %)

Source: G. Jónsson 1998: 34.

Notes: The figures are estimates of the total food supply, based primarily on official agricultural statistics (e.g. livestock, vegetable production) and external trade statistics in addition to various direct evidence on consumption. – Figures for 1784, 1863–1938 show three-year averages.

was transmitted through strong commercial and cultural links with Denmark, but also through the small but influential Danish community living in many of the villages and towns around the country. Food consumption became much more varied and new ingredients were introduced. Danish dishes were adapted to the available food culture, although fewer cereals were used and lamb substituted for pork and poultry.

Around the turn of the twentieth century a new source of change was increasingly felt as urbanization and economic modernization took hold. The emerging coastal towns and villages based their livelihood primarily on fishing, which not only provided an important part of their diet, but paved the way for a new food culture arising from the cash economy. New consumption patterns were fundamental for urban growth, providing rye bread, porridge, margarine, sugar and other cheap imported carbohydrates which soon became staples – in addition to fish – of the urban working class. Even the rural population was shifting its food consumption to some extent from dairy products, meat and blood pudding to cereals and other carbohydrates.

The First World War: Food Provision during an Economic Crisis

Dependence on imported foodstuffs made Icelanders, especially the working class, vulnerable to disruptions in foreign trade. As Figure 16.1 shows, Iceland was importing more than half of its food requirements in 1910 compared to 10 per cent a century earlier. The upheaval caused by the First World War therefore had serious consequences for food provision in Iceland.

Although Iceland was neutral and far from the battlefields of Europe, economic warfare between the Allies and Germany had a huge impact. The economy went into the most serious downturn of the twentieth century between 1914 and 1923, caused primarily by wartime dislocation.[7] Britain had decided early in the war to gain control over Iceland's foreign trade. Due to Britain's status as a sea power and Iceland's dependence on trade with Britain (from where Iceland imported most of its coal and salt), the British government succeeded in forcing Iceland to conclude trade agreements in 1916 and again in 1918 which were very much on its own terms.[8] Fighting between Britain and Germany in the North Atlantic seriously disrupted transport and trade, forcing Iceland to shift most of its trade with Scandinavia and Germany to Britain and, to lesser extent, America.

Iceland was able to secure sufficient food supplies during the first years of the war. However, there were heightened tensions between urban consumers and food producers (farmers and fishing vessel owners) as prices of milk and butter soared and many other foodstuffs became both costlier and scarcer. Meat was a luxury among the urban working class who had to make do with fish and cheap imported foods such as margarine, sugar, coffee and cereals. The government did little to ease the situation except to set up a state-run import company, *Landsverslun*, with the aim of securing some of Iceland's most important provisions: cereals, sugar, coal and salt. By 1918, *Landsverslun* had become the biggest importer in the country.[9] With the escalation of submarine warfare in the North Atlantic in February 1917 the situation rapidly deteriorated, leading to shortages of many imported foods for the rest of the war. Galloping inflation made the food situation critical as poorer sections of society found it almost impossible to afford ordinary foodstuffs.

Wage earners in the biggest towns – Reykjavík, Ísafjörður and Akureyri – were the hardest hit by the food crisis, because they depended on cash income and imported food and fuel. Rising unemployment and substantial cuts in real wages in 1917–18 made their living conditions precarious. Investigation of the situation in Ísafjörður (pop. just under 2,000) in January 1918 showed that 374 families with 1,268 members were destitute and without food and fuel.[10] In Reykjavík where the situation was only a little better, the town council took various measures to ease the

---

7   GDP fell by nearly 18 per cent between 1914 and 1918, see G. Jónsson 2009: 58, 61–4.

8   Jensdóttir 1980: especially chapters II and V, see also Jensdóttir 1986.

9   G. Jónsson 1986.

10   *Lögrétta*, 23 January 1918, 15.

Figure 16.2   Total Energy Value in the Icelandic Diet, 1900–1940 (kcal per capita)

Source: G. Jónsson 1998.

Notes: Figures show three-year averages. The measurement unit in the source is male equivalents but here per capita.

food situation and it started rationing several imported goods in April 1917. Self-sufficiency became a matter of survival for many people who turned increasingly to potato and vegetable growing in and around the towns. Disruptions of trade and record cold temperatures during the winter of 1918 prompted the government to take greater responsibility for food provision, regulating imports, even directly importing essential goods, rationing scarce goods and controlling food prices. In cooperation with the Reykjavík town council it initiated potato growing near Reykjavík, albeit with poor results.[11] Efforts to provide fuel, peat in particular, proved more successful. Various accounts and reports indicate increasing poverty, with a growing number of families in Reykjavík and Ísafjörður unable to support themselves, despite the efforts of the town councils. Many people had to rely on food kitchens and charity organizations over the hardest winter months.

Reliable quantitative data on the war's impact on food consumption is difficult to obtain. The most important national figures are the previously mentioned food supply estimates, which suggest a fall in total energy value in the diet of 17 per cent between 1915 and 1919, as shown in Figure 16.2. Although these are average figures which hide regional and social variations, they strongly indicate a sharp

11   Friðriksson 1994: 8–9.

decrease in the food supply during the war. This was largely due to a reduction in cereal imports.

The social upheaval caused by the war made food a political focus point and marked the beginning of modern consumer politics in Iceland.[12] Housewives protested against rising food prices in August 1914, the first food-related protests in Iceland. Consumers and workers founded co-operatives and the tension between the urban and rural population gave rise to the formation of pressure groups and political parties based on class and social status instead of the stance over the constitutional ties with Denmark.

The Interwar Years: The Rise of Agrarianism

The production, distribution and consumption of food can be regarded as the underlying source of most major political disputes in twentieth century Iceland. Icelanders were first and foremost food producers as more than a third of the workforce was engaged in food production and most exports were foodstuffs, primarily fish products of various kinds.

Modernization and urban growth created friction between the urban and rural population. At the core of the debate on food politics were tensions between two developmental paths which could be termed export-oriented capitalism and agrarianism. Supporters of the former saw export specialization, based on the country's rich fishing stocks, as the most important growth strategy for Iceland. They advocated a liberal trade policy and a fairly passive adjustment to international markets. The path of economic modernization lay through division of labour, efficiency and innovation-based technology. Supporters of the latter course favoured 'economic nationalism', emphasizing self-sufficiency and import substitution to support domestic industries. They advocated small-scale production in which farming life was at the core of society and believed that agriculture had great growth potential if given the right conditions to develop. The optimists even believed that Iceland could become a major exporter of meat and butter. Agriculture should be organized as family operations while related industries would be run by co-operatives, thus aiming at improving the capabilities of the 'collective entrepreneur'.[13]

The political cleavage crystallized in the two major power blocs dominating Icelandic politics during the interwar period. On the one side were the liberal business class and fishing vessel owners, united in a strong party with the founding of the Independence Party in 1929. On the other were the agrarian interests, organized in the Progressive Party in 1916, a leading party in Icelandic politics until the Second World War and closely linked to the co-operative movement. The core of the third and weakest political bloc was the trade union movement.

---

12   G. Jónsson 2007: 77–8.
13   See e.g. Ásgeirsson 1988. These ideas are akin to Chayanov 1986.

It established the Social Democratic Party in 1916, which split with the founding of the Communist Party in 1930. The left was also weakened by an electoral system that greatly favoured rural constituencies at the cost of towns. This gave the farming community and their representatives much greater political leverage than their size and economic significance warranted.[14]

Disputes about economic and social organization manifested themselves in conflicts between urban and rural interests. Until the First World War liberal politics were gaining strength and Iceland consistently followed a free trade policy with an emphasis on exploiting the fishing grounds and creating an economic infrastructure that facilitated fish exports.[15] The balance of power shifted in favour of the agrarian interests during the interwar period, leading to significantly increased public support for agriculture through cultivation schemes and development programmes. The Progressive Party managed to get two important laws adopted by parliament in 1934 and 1935 with the help of the Social Democrats. These introduced regulation of the mutton and milk markets. Competition was all but eliminated and marketing boards were established to regulate supply and set prices without regard for efficiency or proximity to markets. Most importantly, with the advent of the Great Depression, Iceland was forced to abandon its long-standing liberal trade policy and put up trade barriers in 1931. Comprehensive import-substitution policy was adopted in 1934, banning or severely restricting the imports of most foodstuffs. The meat and milk markets were regulated with the aim of ensuring farmers minimum prices for their products and plans were drawn up to substitute imported foods, such as margarine, with domestically produced butter. The cost differences were to be subsidized by the taxpayer. The political outcome was a kind of agrarian-based populism.

Interestingly, in spite of the Social Democrats being in government from 1934 to 1942, the interests of consumers were not given high priority in public policy. It was the smallest party in successive coalition governments and some of its leaders were influenced by the ideology of agrarianism and the ideal of the independent small producer. Working-class families, especially in the smaller towns and villages, tried to secure their food supplies by growing vegetables and taking up small-scale farming with perhaps a cow and several sheep. This type of living, however, was not common in many coastal towns such as Ísafjörður in the West Fjords, Siglufjörður in North Iceland and Seyðisfjörður in the East Fjords, where access to farmland was limited. As a result these towns had a more proletarian character.

Supplying the fast-growing urban areas with food was a major challenge at a time when the economic infrastructure was undeveloped. Even in Reykjavík, where the largest market and the most advanced technology existed, the situation was critical. Road transport (no railways existed) to major agricultural districts to the north and east of Reykjavík only became possible in the late 1920s and it took

---

14   Ásgeirsson 1988, Ö.D. Jónsson 1996.
15   G. Jónsson 2004, 153–58, Ásgeirsson 1988: 13–42.

up to two days to transport products from dairies in South Iceland to Reykjavík – the distance was only 60 km but the road had to go through a mountain pass.[16]

The contraction of foreign trade further strengthened the agrarian vis-à-vis the capitalist interests in Iceland. Salt fish had been the most important export article, but the market was rapidly declining, especially with the outbreak of the Spanish Civil War, since Spain had been the biggest importer of salt fish. There was, therefore, a combination of factors working against a more efficient, market-based organization of the economy during the 1930s which eventually reduced production capacity. Even in the sphere of consumption the farmers' organizations were on the rise and they exerted ever greater influence through wholesale and retail cooperatives which primarily promoted the interests of the farming community. Consumer cooperatives were, however, few and of little economic significance.

What implications did these broad political and economic changes have on food consumption and consumer policy in Iceland? Firstly, the overall energy intake corresponds well with the country's changing economic fortunes during the interwar period. It rose significantly during the 1920s, then stagnated after 1930 and even declined slightly in the late 1930s. Looking at the diet and individual foodstuffs, the shift from animal to vegetable foodstuffs was still underway in the 1930s, albeit at a slower pace. Consumption of butter, fish and, especially, mutton declined while consumption of sugar, margarine and potatoes rose. Two exceptions to the trend are notable in the 1930s: there was a steady increase in consumption of dairy products and a drop in cereals consumption after 1936.

The protective tariffs set up to shelter domestic production severely restricted imports of various types of foods that Icelanders were accustomed to, setting a lasting mark on food culture. Imports of meat, fish, eggs, biscuits and cakes were severely restricted or banned; all of these products had been consumed in considerable quantities before the crisis. In 1932–33, an import ban was introduced on margarine, cheese and condensed milk, the last product having been very popular among the urban poor who could not afford domestically produced milk. Imports of fresh and dried fruits were also banned. As a result of these measures the Icelandic diet became much more monotonous and the trend towards greater variety and a more broadly based diet, which had been one of the most important features of the modernization process, ceased for a while. In one area of food culture, however, growth and creativity prevailed. Imports of cheap carbohydrates like sugar and starch were unrestricted and as a result sugar consumption more than doubled during the early decades of the twentieth century, rising from 20 kg per capita around 1900 to 44 kg just before the Second World War. This placed Iceland among the highest sugar consumers in Europe.[17] Sweet-tasting foods included puddings and an enormous variety of cakes and biscuits became popular. The modernized Icelandic diet had a sweet taste.

16　Jónsson, Gudmundsson and Lýðsson 1989.
17　*Icelandic Historical Statistics*: 657.

The Second World War: Affluence but limited Americanization

Iceland's fate during the Second World War was very different from that of the First World War. Iceland was one of very few countries in Europe to experience high economic growth during the war, a complete transformation from the bleak years of the depression. The Second World War fundamentally altered the fortunes of Iceland, turning it into a booming economy with one of the highest growth rates in Europe. By the end of the war, the country had changed from one of the poorest to one of the richest of Western Europe, rising to fourth place in the European league in terms of GDP per capita, preceded only by Britain, Sweden and Switzerland.[18] The war brought unparalleled economic expansion and radical improvement in living standards. The export industries, especially the fishing sector, profited greatly from increased demand abroad, mainly from Britain. The occupation of Iceland in 1940 by British and later US forces had an even bigger impact as tens of thousands of troops swamped the tiny Icelandic population of only 120,000.

In contrast to the First World War, the Allies now pursued a much friendlier policy towards Iceland, offering the government favourable trade agreements in order to promote better political relations with the occupied nation. Foreign trade shifted mainly towards Britain and the United States through trade agreements which were on the whole highly favourable for Iceland. The British authorities were concerned about Iceland's imports for several reasons. Although they wished to maintain Icelandic purchases of British goods, they were unhappy at uncontrolled exports, especially of non-essential goods, to Iceland. More generally, the British government believed that Iceland's food consumption was excessive and in early 1941 came to the conclusion that 'total control over the intake of foodstuffs into Iceland' was overdue because consumption levels were much too high both in light of wartime conditions and Icelandic pre-war levels.[19] Being in a position to dictate Iceland's imports policy through its control of Icelandic trade and shipping, Britain eventually persuaded the Icelandic government in 1941 to accept an import programme. This was based on the principle that Icelandic imports should be limited to average pre-war levels 'though increases in certain categories would be permitted where we are satisfied that they are justified'.[20] In accordance with the programme, imports of many food classes such as cereals, sugar, fruit, tea and coffee and oil and fats were to be reduced.

Import statistics indicate, however, that the British policy was not successful. The opportunities brought about by the war meant that Iceland had vastly greater resources to escape from poverty and malnutrition. Consumers' purchasing power

---

18 G. Jónsson 2004: 146–8.

19 The National Archives, Kew, (TNA), MAF 83/161, Minute ECA 237/5, R.N. Chapman 30 April 1941; Cable from Ministry of Economic Warfare, 25 June 1941.

20 TNA, T 160/213, F 16978: 2. Note of a meeting to discuss policy in relation to Iceland, 7 June 1941.

had grown significantly, although purchases were limited by wartime circumstances. One of the reasons for the limited success of the British in reducing Icelandic imports was the fact that Iceland was shifting its import trade from Britain to the United States as early as 1941, resulting in reduced British control of the Icelandic trade. The US government was party to the agreement about the Icelandic import programme from 1942 onwards, but it did not pursue as strict a policy as its British counterpart. As a result, Icelandic importers were able to increase significantly their purchases of various consumer goods, including foodstuffs, from the United States. For example, the real value of total cereal imports doubled between 1939 and 1945, and there was a 15-fold increase in imports of fruit and nuts.[21]

Even so, shortages of many food articles were frequent during the war. As early as September 1939 the Icelandic government started rationing many foodstuffs and necessities, including bread and cereals, coffee and sugar. Public measures were taken to regulate food supplies and consumption, not only to meet the requirements of the annual import programmes, but also to keep control of food prices and distribution between regions. However, import restrictions were not severe and rationing was not as strict as during the First World War.[22]

The Icelandic diet seems to have changed more in terms of quantity than quality during the war. The influence of British and American food cultures as a result of the occupation has frequently been cited as causing a major change in consumption habits, but this claim is mostly based on anecdotal evidence. Of course, many foods were now imported from America instead of Denmark, Germany or Britain, but these were mostly the same staple goods: sugar, rice, maize, flour and other cereals. Fresh and dried fruits were perhaps the greatest novelties from America. The troops brought with them spam, chocolate, chewing gum and soft drinks like Coca-Cola, but these were all superfluous foods that did not make any significant impact on the daily fare of ordinary people. And although fish and chips shops set up in Reykjavík with the advent of British troops, the Icelandic cuisine did not change much. Probably the greatest change was that most Icelanders could now afford to be reasonably well fed by consuming their traditional foods: fish, dairy products and mutton. Ordinary people could now afford to eat mutton twice a week. Canned and dried fruits, which had disappeared from the shelves in the shops during the depression, appeared again and were eaten in considerable quantities. Alcohol consumption also increased significantly.

The war brought unprecedented wealth to Icelandic society, fostering consumerism and new lifestyles. However, the indications are that the occupation forces made little profound and lasting impact on the food habits of Icelanders. The Icelandic political economy, dominated by productionist interests and economic nationalism, remained intact for some time yet.

---

21   *Icelandic Trade Statistics* 1939–1945.
22   A.K. Jónsson 1969.

Reading Social Transformation from Food Recipes

One way to acquire insight into the impact of the severe shocks of the two world wars on the diet is to apply a micro approach by looking at how the needs and preferences of households were articulated through cookbooks. At first sight, it might seem a somewhat frivolous approach to difficult and troubled times, but we take the view that cookbooks reflect important elements in the food culture, even though they often are intended to educate and change prevalent food habits. We examine the works of two of the most influential home economics teachers of the period, Jóninna Sigurðardóttir and Helga Sigurðardóttir. The former was a good representative of food styles of 1910–30 and the latter a well-known educator, who was determined to rescue her country from the tradition-bound and narrow-minded food habits of agrarian Iceland.

Jóninna Sigurðardóttir studied in Denmark at several home economics schools for three years. She established her own school upon her return to Iceland, after working as a migrant teacher educating housewives.[23] Jóninna published several books which became widely popular, especially the first one, *Ný Matreiðslubók fyrir fátæka og ríka* [*New Cookbook for the Poor and the Rich*], which was reprinted seven times between 1915 and 1945.[24] A noteworthy feature of the first edition is a cost–calorie ratio table, showing the calorie price, i.e. calories per penny (or *öre* in this case) of all the main ingredients used in the recipes. It does not come as a surprise that eggs are the most expensive food (only 14 calories per penny in cost), beef came second (20 calories), followed by cheese (30 calories) and milk (36 calories). The cheapest food was herring (200 calories) followed by maize (128 calories) and oats (115 calories); sugar is a middle-range food, providing 80 calories per penny spent.[25]

Nearly all of the recipes in Jóninna's cookbook are adaptions of Danish cooking and the book is influenced by a Danish cookbook, *Ökonomisk kogebok*, by the Danish scientist, Mikkel Hindhede.[26] The generous use of butter in most of the recipes either indicates a rather limited adaptation to Icelandic circumstances or simply wishful thinking considering that butter was both difficult to get and quite expensive. In the book's preface, written by the district physician, a very 'modern' view is expressed. Icelanders' insatiable demand for fish and meat means a high intake of proteins, which are seen as 'unclean' as they leave behind undesirable residues which are difficult to digest. By contrast, carbohydrates and fats are pure energy which can be directly compared to any other kind of energy; the metaphor used is petrol fuelling a car.

In the 1945 edition of the book the reference to rich and poor has disappeared, but the recipes have not changed much. Danish food culture is still dominant and,

---

23 Einarsdóttir 1986.
24 Ibid.
25 J. Sigurðardóttir 1915: 6.
26 See Chapter 15 by Svend Skafte Overgaard.

if anything, even more pronounced than in the first edition. Herring was now an acceptable food for the better off as *Islandsild* (Icelandic herring) had become highly valued in Danish food culture. One of the big differences between the food cultures of the two countries was that the Icelandic one lacked several key ingredients of the Danish fare: corn-fed pigs, poultry and eggs. The sparse use of flour, especially wheat, and eggs influenced the Icelandic way of making cakes, cookies and pastries. Instead of butter and lard, Icelanders resorted mainly to margarine except on special occasions. Jóninna's cookbook reflects the rise of a carbohydrate-based food culture. In the 210 page long book only 10 pages are devoted to meat and fish recipes. The rest of the book contains a vast number of recipes with vegetables (mainly potatoes), endless variations of soups and, more importantly, cakes and biscuits of all sorts.

Helga Sigurðardóttir was also a Danish educated home economics teacher. She moved to Reykjavík and became one of the most influential campaigners of 'modern' cooking, emphasizing the use of fresh ingredients and a more varied diet.[27] She was influenced by Danish food traditions, but she also advocated much greater use of domestically grown berries and vegetables of various kinds. Due to the short summer season she took a great interest in improving preservation methods.[28] Helga's ideas were closely linked to a greater interest in vegetable growing during the interwar period when especially townspeople started growing their own vegetables in home gardens or plots outside town. This was certainly the case in Reykjavík and in nearby Hveragerði geothermal energy was used for heating greenhouses where potatoes, turnips, carrots and cabbage were grown. Vegetables had long been regarded as an inferior substitute for 'proper' food, but in hard times, as during the 1930s, potatoes and other vegetables saved many families from hunger.[29] This new and increasingly important source of nourishment may partly be seen as a response by townspeople to protectionist trade policies, taking food production in their own hands since they were denied cheap imported foods.

Conclusion

Stronger links with international trade and economic modernization gradually transformed the Icelandic diet in the course of the nineteenth century. In contrast to most European countries, the Icelandic diet was moving away from animal foods to cheap imported carbohydrates, especially among the rising urban working class. The emerging food culture arose from the cash economy and was highly dependent on uninterrupted foreign trade. The upheaval and dislocation caused by two world wars totally upset Icelandic foreign trade and the economy in

---

27   Her *Lærið að matbúa* [Learn How to Cook] and *Matur og drykkur* [Food and Drink] became standard cookbooks in Icelandic households for decades.

28   H. Sigurðardóttir 1932, H. Sigurðardóttir 1940.

29   Sigurðsson 1995.

general, but the wars' effects were very different. In the First World War Iceland experienced a serious and prolonged economic crisis, trade disruptions and food shortages, which created a critical situation among the urban poor. The Second World War, by contrast, brought unprecedented wealth to Icelanders who were able to afford more and more varied food than before. However, the composition of the diet did not change in any fundamental way, despite the influx of tens of thousands of Allied troops.

Although the wars disrupted the 'modernization' of the national diet, they did not seem to have had a lasting impact on its composition. The important changes in the first half of the twentieth century can be traced primarily to the political economy of Iceland, especially the strong position of agrarian interests and the abandonment of free trade, which shifted consumption patterns partly back to domestically produced foods. In one area, however, the ordinary consumer was allowed to take advantage of cheap imported foods, and that was in the consumption of sweet-tasting foods. Icelanders became one of Europe's biggest sugar consumers, with a huge appetite for sweet foods and soft drinks. No wonder that the Icelandic cuisine was at its most creative when it came to sumptuous cakes, biscuits and pastries.

References

Ásgeirsson, Ó. 1988. *Iðnbylting hugarfarsins. Átök um atvinnuþróun á Íslandi 1900–1940.* [The Zeitgeist of Industrialization. Conflicts over the Economic Development of Iceland] Sagnfræðirannsóknir 9, Reykjavík: Sagnfræðistofnun.

Chayanov, A.V. 1986. *On The Theory of Peasant Economy*, edited by D. Thorner, B. Kerblay and R.E.F. Smith. Madison, WI: University of Wisconsin Press.

Einarsdóttir, B. 1986. *Bókvit í aska. Úr ævi og starfi íslenskra kvenna.* [The Lives and Work of Icelandic Women] III, Reykjavík: Bókrún.

Friðriksson, G. 1994. *Saga Reykjavíkur 1870–1940.* [The History of Reykjavík, 1870–1940] II, Reykjavík: Iðunn.

Gísladóttir, H. 1994. The use of whey in Icelandic households, in *Milk and Milk Products from Medieval to Modern Times: Proceedings of the Ninth International Conference on Ethnological Food Research*, Ireland, 1992, edited by P. Lysaght. Edinburgh: Canongate Academic, 123–9.

Gísladóttir, H. 1999. *Íslensk matarhefð.* [Icelandic Food Culture] Reykjavík: Mál og menning.

Grigg, D. 1995. The nutritional transition in Western Europe. *Journal of Historical Geography* 22(1), 247–61.

Icelandic Historical Statistics. 1997. Ed. by Guðmundur Jónsson and Magnús S. Magnússon, Reykjavík: Statistics Iceland.

Icelandic Trade Statistics. 1939–1945. Reykjavík: Statistics Iceland.

Jensdóttir, S.B. 1986. Anglo-Icelandic Relations during the First World War. New York: Garland.

Jensdóttir, S.B. 1980. *Ísland á brezku valdsvæði 1914–1918.* [Iceland within the British Sphere of Influence, 1914–1918] Sagnfræðirannsóknir 6, Reykjavík: Sagnfræðistofnun.

Jónsson, A.K. 1969. *Stjórnarráð Íslands 1904–1964.* [Government Offices of Iceland 1904–1964] II, Reykjavík: Sögufélag.

Jónsson, G. 1986. Baráttan um Landsverslun 1914–1927. [The dispute over the state trading Company] in *Landshagir. Þættir úr íslenzkri atvinnusögu,* edited by H. Þorleifsson. Reykjavík: Landsbanki Íslands, 115–38.

Jónsson, G. 1998. Changes in food consumption in Iceland ca. 1770–1940. *Scandinavian Economic History Review,* XLVI(1), 24–41.

Jónsson, G. 2004. The transformation of the Icelandic economy: industrialisation and economic growth, 1870–1950, in *Exploring Economic Growth: Essays in Measurement and Analysis. A Festschritft for Riitta Hjerppe on the 60th Birthday,* edited by S. Heikkinen and J. L. van Zanden. Amsterdam: Aksant, 146–8.

Jónsson, G. 2007. Hvenær varð neysluþjóðfélagið til? [When did Consumer Society Begin to Emerge?] in *Þriðja íslenska söguþingið 18.–21. maí 2006 Ráðstefnurit,* edited by B. Eyþórsson og H. Lárusson. Reykjavík: Aðstandendur þriðja íslenska söguþingsins, 69–78.

Jónsson, G. 2009. Efnahagskreppur á Íslandi 1870–2000. [Economic crises in Iceland] *Saga* XLVII(1), 45–74.

Jónsson, Ö.D. 1994. Nature as a demanding ally. *Pesto Papers,* Aalborg: The Aalborg University.

Jónsson, Ö.D. 1996. The geopolitics of fish: The case of North Atlantic, in *Fisheries Management in Crisis: A Social Science Perspective,* edited by D. Symes and K. Crean. Oxford: Fishing News Books, 187–94.

Jónsson, Ö.D. 2005. The creative fight for survival. Food consumption in Iceland – classical tale of rags to riches, in *Celebrating Europe at the Dinner Table: Food, Culture and Diversity,* edited by D. Goldstein. Strasbourg: Council of Europe Publication, 217–27.

Jónsson, S., Guðmundsson, J., and Lýðsson, P. 1989. *Flóabúið – Saga Mjólkurbús Flóamanna í 60 ár.* [The 60 Years History of the Flóabú Dairy] Selfoss: Mjólkurbú Flóamanna.

Pendergrast, M. 2000. *For God, Country and Coca-Cola: The Definitive History of The Great American Soft Drink and The Company that Makes It.* New York: Basic Books.

Sigurðardóttir, H. 1932. *150 jurtaréttir.* [150 Vegetarian Recipes] Reykjavík.

Sigurðardóttir, H. 1940. *Grænmeti og ber allt árið. 300 nýjir jurtaréttir.* [Vegetables and Berries All Year Round. 300 New Vegetarian Recipes] Reykjavík: Ísafoldarprentsmiðja.

Sigurðardóttir, H. 1934. *Lærið að matbúa. Matreiðslubók og ágrip af næringarfræði.* [Learn How to Cook. A Cookbook and a Synopsis of Nutrition] Reykjavík.

Sigurðardóttir, H. 1947. *Matur og drykkur.* [Food and Drink] Reykjavík: Ísafold.

Sigurðardóttir, J. 1915. *Ný matreiðslubók fyrir fátæka og ríka.* [New Cookbook for the Poor and Rich] Akureyri.

Sigurðsson, H. 1995. *Hallir gróðurs háar rísa. Saga ylræktar á Íslandi á 20. öld.* [The History of Greenhouse Cultivation in Iceland in the 20th Century] Reykjavík: Samband garðyrkjubænda.

Chapter 17

Horsemeat in France: A Food Item that Appeared during the War of 1870 and Disappeared after the Second World War

Alain Drouard

Introduction

Hippophagy is an old practice which can be found in prehistoric times, but religious prohibition, royal decrees and prejudices prevented horsemeat consumption in France and Europe generally for centuries. It was not until the second half of the nineteenth century that horsemeat consumption became authorized and that horsemeat became a 'popular' food, accessible to most. Over the course of a century, between the 1870s and 1975, consumption increased, reaching a peak after the Second World War. Considered the beef of the poor, horsemeat was sold at horsemeat butchers' shops which multiplied in Paris and the provinces. At the same time, the trade was getting organized and the state regulated and controlled horsemeat production and consumption. The French were among the leading consumers of horsemeat, along with the Belgians. After this long thriving period, a crisis occurred as motorized vehicles replaced horses, reducing the supply and raising prices. The consumption of horsemeat, which declined over the past 25 years, is minimal today.

Religious Prohibition

Judaism proscribed the consumption of horsemeat and Christianity, in its turn, established the same prohibition.[1] Isidore Geoffroy Saint-Hilaire (1805–1861), Etienne Geoffroy Saint-Hilaire's son and Professor at the Museum of Natural History, one of the principal defenders of horsemeat in France, explained in 1868 that,

> Just like some Asian nations still do it nowadays, the former peoples of the North and the center of Europe, the Suèves, the Vandals, the Celts, and the

---

1 Deuteronomy 14: 4–5, mentions ten authorized animals, ox, sheep, goat, stag, gazelle, deer, ibex, antelope, buffalo and roe-deer. The horse is not mentioned.

Germans used to feed on horse milk and blood. This use which still existed in the 8th century of our era was connected to certain religious practices. During their festivals the sacrifice of a horse was followed by a meal where the flesh of the victim was eaten. The persistence of this pagan custom was an obstacle to the propagation of Christianity. Therefore, Pope Gregory III ordered Saint Boniface to prevent this act of idolatry by all means and to declare horsemeat abject and revolting and Zachary had to renew the prohibition.[2]

As Wagner demonstrates, the Pope's letter does not refer so much to animal sacrifice as to the impurity of horsemeat,

> You still tell me that some people eat wild horsemeat, and that many also eat domestic horsemeat. From now on, very holy brother, you should not tolerate these practices, but on the contrary completely prohibit them by all means possible, with the assistance of Christ, and impose a deserved penitence, for it is impure and detestable.[3]

Religious prohibition lost its influence over the centuries, but usage, custom or prejudice opposed the consumption of horsemeat. According to Madeleine Ferrieres it was difficult to eat a 'noble' animal, one of the best companions of man.[4] As Buffon, writing in 1753, put it, 'The most noble conquest man ever made is this proud and fiery horse that shares with him the war's fatigue and the fight's glory'.[5] Health or economic arguments should also be mentioned. Butchers did not want to lose the monopoly on the sale of meat. Many edicts and orders made at their request (in 1735, 1739, 1762, 1780, 1802 and 1811) repeated the prohibition of the sale of horsemeat, which was then conducted by peddlers or illicit dealers.

Regardless, due to necessity during times of famine, war or siege, people would eat horsemeat. It can be noted in military annals that horsemeat soups were distributed by Baron Larrey to the soldiers of the Great Army after the battle of Eylau,

> The muscular horse flesh, especially that of the hindquarters, can be used in the preparation of a soup, especially if you add a certain amount of bacon; it can even be used to be grilled and as beef à la mode, with the suitable seasoning...
>
> During the battle of Eylau, for the first twenty four hours, I still had to feed my wounded soldiers with horse flesh made as a soup or as beef à la mode; but as the seasoning was not scarce in this circumstance, the wounded soldiers could barely tell the difference between this meat and beef.[6]

---

2   From an article published after his death in *Le Moniteur universel*.
3   Wagner 2005: 468.
4   Ferrières 2007: 445.
5   Buffon 1753: 174.
6   Parent-Duchatelet 1832: 120, quoting Baron Larrey.

Horsemeat was therefore eaten before the late nineteenth century. Its consumption was disguised or it was eaten in the form of sausages. Isidore Geoffroy Saint-Hilaire mentions the consumption of horsemeat under other names; 'most often' it appeared 'under the name of beef or as venison in restaurants (some of the highest orders) without consumers suspecting fraud or complaining'.[7]

A Skilful Promotion Campaign

Prohibition was eventually lifted after hygienists, scientists, doctors and veterinarians campaigned to improve the condition of the working class by providing them with a food item that was until then reserved for the rich, namely meat. For Bizet, a horsemeat advocate, the key question was to increase the supply of cheap meat, thereby making available to the poor as well as the rich a food of first necessity. The main issue was, 'How to lower meat's prices and to increase its consumption'.[8] Parmentier, Larrey, Geoffroy Saint-Hilaire, Huzard, Decroix, Renaut, Leblanc, Goubaux, de Quatrefages and Piétrement lectured and published extensively with the aim to convince people of the benefits of horsemeat. Many called upon the example of foreign countries like Belgium,[9] Germany,[10] and Denmark (Delavigne 1999), which preceded France by about 15 years.

Isidore Geoffroy Saint-Hilaire was the most fervent advocate of horsemeat (1856a). In *De l'usage alimentaire de la viande de cheval (1856 b)*, a brief summary of his book which was published the same year, he precisely defined his objective:

> The reasoning that I can give is three-fold: horsemeat is healthy; it tastes good; there is enough of it so that it can very usefully become a part of people's diet.
>
> There is in the use of horsemeat an important resource for the diet of the working classes, the most important one (even though it is not enough yet) we can resort to give them what they lack today above all: the food item by excellence, meat.
>
> A singular social anomaly which we will be astonished one day to have undergone even for such a long time! There are millions of French people who do not eat meat; they eat meat six times, twice, once a year! And considering this misery, each month millions of kilos of good meat are given to the industry

---

7   Geoffroy Saint-Hilaire 1856a: 457.

8   Bizet 1847: 347.

9   The consumption of horsemeat increased in Belgium from 1860 onwards, consumed mostly as delicatessen such as horsemeat sausages called 'dry sausage of Boulogne' or cervelat. In 1896, Belgium had a total of 13 establishments which manufactured dry sausage of Boulogne and cervelat.

10   Germany was the first country to hold banquets serving horsemeat beginning in the early 1840s.

for very secondary uses, given to pigs or dogs, or even thrown to the streets all
over France!

The defenders of horsemeat also organized meals entirely made of horsemeat.
The first horsemeat banquet took place on 1 December 1855 in Alfort under
the direction of the veterinary school director, Renault. On 6 February 1856 a
second banquet was held, this time in Paris, in the *salons* of the Grand Hotel.
The 132 guests who included veterinarians, scientists like Quatrefages, bankers,
industrialists, notaries, journalists and agronomists were all well known and
influential people.

The veterinarian Emile Decroix (1821–1901) continued the campaign after
Geoffroy Saint Hilaire died before achieving his goal. In 1864, he founded
the Committee for the Propagation of Horsemeat. It was chaired by Blatin, a
veterinarian. To popularize its pursuits, the committee had a horse killed and its
meat distributed to poor families every week. Decroix, a remarkable propagandist,
did not hesitate to take risks to achieve his goal. He ate horsemeat coming from
animals infected with equine influenza and 'any other disease' to show that it was
harmless. He declared,

> I will share a certain number of these experiments.... The ones I will mention
> will be enough, I think, to justify this assertion sustained for the first time at the
> *Société d'acclimatation* during the siege of Paris namely:
> 1. That horsemeat is good for man's diet.
> 2. That one can use the cooked flesh of an animal that died of any disease
> without any risk.[11]

In 1864 a first request to open a 'horsemeat' butcher's shop was filed with
the Minister for Agriculture, Trade and Public Affairs. The Council of Public
Health gave a favorable report. Under these conditions, prohibition no longer
made any sense. In Paris, the ordinance of the Prefect of police of 9 June 1866
authorized the sale of horsemeat for human consumption. This authorization was
accompanied by many prohibitions. Horses were not to be shot down any more
in 'fields of squaring',[12] but killed in slaughterhouses 'specifically authorized for
this purpose and located in the district of the Police headquarter'. Slaughtering
was to be done 'in the presence of a veterinarian or an inspector commissioned
by the Prefect of police'.[13] Finally, horsemeat was to be sold in special butcher's

---

11   Decroix 1885: 7.

12   Only wounded or sick animals were to be shot down in the fields of squaring and
their consumption was prohibited.

13   Police decree of 9 June 1866, regarding the selling of horsemeat for human
consumption. Article 3 reads, 'Transportation, and sale of horsemeat coming from
slaughterhouses other than those indicated in the preceding article are prohibited in Paris
and the rural villages under our jurisdiction'.

Figure 17.1   The First Horsemeat Butcher's Shop in Paris

Source: *Le Monde illustré*, 22 September 1866

shops which were different from other butcher's shops and easily identifiable; 'The stalls assigned to the flow of horsemeat will be indicated to the public by a sign in capital letters announcing their specialty'. The first horsemeat butcher's shop was opened in Paris on 9 July 1866 at 3, boulevard d'Italie, (later Place d'Italie) by a man called Antoine who received a bonus from Emile Decroix.[14] It was located in a working-class district and very close to the horse market of the boulevard de l'Hôpital. The second and third butcher's shops also opened in working-class districts, at 101 rue de Clichy in September and at 10 rue Desnoyers in Belleville in October 1866.[15] Were the theoretical arguments of the proponents of horsemeat enough to convince the population?

---

14   Outside Paris, the first horsemeat butcher's shop to open was in Nancy. In the work dedicated to the Fiftieth anniversary of Vaugirard's slaughterhouses (1954: 8), a M. Thouin is mentioned as the first equine butcher.

15   Fierro 1996: 724.

The War of 1870 and the Siege of Paris

Emile Decroix was right to stress that the take-off point in the consumption of horsemeat was the siege of Paris by the Prussians in 1870–71. Until this date the figures remain modest.[16] In 1871 Molinari provided one of the most interesting testimonies on the beginnings of horsemeat consumption in Paris;

> People bravely started eating horsemeat; the wealthy classes set the example, and little by little the popular disgust for this uncommon food was overcome. Some servants of good houses still refuse to touch the filet's leftovers or the rib steak which was the main dish of the masters' dinner, but the number of these recalcitrant people diminishes day by day, and the convinced horsemeat eaters are not far from believing that the introduction of horsemeat into the public diet could compensate, up to a certain point, for the evils of the siege and the disasters of the invasion.[17]

Before the siege the number of horses in the capital was estimated at 100,000. During the first three months of the siege 30,000 were killed to feed the Parisian population. In the following months, about 30,000 more horses were killed. Parisians, the bourgeoisie as well as the working class, ate horsemeat in preference to the animals of the botanical garden or cats and rats.

On 11 September 1870, soon after having restored the tax on butcher's meat, the government regulated the sale and marketing of horsemeat with effect from October 1870. A decree of the Minister for Agriculture and Trade on 7 October stipulated:

> Article 1: The horses intended for food will have to be sold on Mondays, Wednesdays and Fridays of each week, from 8 to 11am, on the Horse Market.
> Article 2: Only the horses declared healthy by the veterinary service of inspection of the market will be able to be sold for consumption. These horses will be able to be killed only in the slaughter-houses.
> Article 3: The horses bought by the State will be weighed alive on the market scale and will be paid for in cash at the maximum price of 40 cents per kilo.
> Article 4: On the stalls authorized to sell horsemeat, the selling price of the aforementioned meat is fixed as follows:
> Sirloin, section, rump, topside, silverside: 1.40 franc per kilo
> All other pieces: 80 cents per kilo....

---

16   The figures stood at 906 horses in 1866; 2,152 horses including 59 donkeys and 24 mules in 1867; and 2,758 horses in 1868.

17   Molinari 1871: 114–15.

Horsemeat was less expensive than mutton and beef, the prices of which had been fixed by an earlier decree.[18] Later decrees by district mayors created municipal butcher shops which sold the meat at the price of the tax. They also instituted a census of the population and introduced ration cards.[19] At same time, in the provinces the war also led to the regulation and public control on horsemeat. This is illustrated by a decree of 23 September 1870 ordered by the division general, F. Higher Coffinières, superior commander in Metz;

> His Excellence Mr. Marshal Bazaine, commander-in-chief of the army agreed to yield to the city the number of horses necessary for public food. This gift is made under such conditions that the fixed-prices in the decree of September 15 will be lowered, as indicated below, starting on the 25th of this month.
> Horsemeat, low quality 10 cts per kilo
> Horsemeat, medium quality 50 cts per kilo
> Horsemeat, premium quality (except sirloin) 1 franc.[20]

Apart from its price, which was to make it a popular meat, horsemeat was also promoted as a healthy meat. In any case, the government of the Third Republic provided guarantees to consumers by ensuring the control of the animals destined for human consumption.[21] The town of Saint-Denis encouraged the sale of horsemeat, specifying that

---

18   Decree of 6 October 1870 fixed the price for beef and mutton:
Beef:
1st category: 2.10 f per kilo (section, rump, topside, silverside, sirloin)
2nd category: 1.70 f per kilo (chuck, ribs, heel of collar, sirloin flap, kidneys)
3rd category: 1.30 per kilo (collar, teat, topsides, rib dishes, sirloin, cheeks). The tenderloin and the sirloin are taxed at 3 f per kilo.
Mutton:
1st category: legs and racks: 1.80 f per kilo
2nd category: shoulders 1,30 f per kilo
3rd category: brisket, neck, ribs: 1.10 f per kilo.
19   Decree of 11 October 1870 by the mayor of the 11th district; decree of 9 October 1870 by the mayor of the 14th district; regulations of the municipal butcher's shops of the 6th district; notice on the rationing by the mayor of the 5th district dated 14 October 1870.
20   *Les Murailles d'Alsace-Lorraine*, 1874: 54. In Metz and Paris, the siege of the city resulted in the introduction of rationing and the requisitioning of horses (decree of 24 October 1870).
21   The decree of the Ministry for Agriculture and Trade dated 20 October 1870 reminded people that horses were to be shot down after veterinary inspection only in the slaughter-houses. It also specified that these horses 'will be marked by a scarlet letter on the left hip' and that 'the entry into these slaughter-houses is formally prohibited to horses not carrying the mark of the market inspection'.

The Council, after gathering information on the hygiene of a horsemeat diet, decided that this meat could be sold freely in Saint-Denis, under the condition of making a preliminary declaration to the Town hall, and to comply with the laws, rules and local decrees regarding the trade of butcher's shops, but also under the express condition that the meat offered to consumption will come from live horses shot down in the city after being subjected to the visit of a doctor – veterinarians....[22]

Mutton and beef became increasingly scarce due to the siege, but paradoxically it actually contributed to the promotion of horsemeat. On 21 October 1870 the Minister of Trade, J. Magnin, informed the mayor of the 2nd district that,

The reduction in the number of oxen and sheep supplied to the city of Paris and the need to ensure for a longer time the means of National defense force me to restrict the amount of kilos of fresh meat that I had previously allocated to your district in the daily distribution of slaughter-houses...

For the population to be less affected in its usual consumption, it seems to me that you could compensate for the deficit in beef and mutton with horsemeat, which in all fairness is well accepted today by many people, and provides as healthy and strengthening a meat as beef.

You know that selling horsemeat is free and has no other obstacle than tax; it would thus be easy for you to strike a deal with the butcher's who sell this kind of meat to provide to Municipal Butcher's shops the equivalent of beef that the current measure is going to take away from your district.[23]

Meat from horses and mules was taxed after 8 November, but this did not prevent butchers from selling it at higher prices and the state was forced to intervene again to fix the price of horsemeat.[24] Because of the siege and the increasing scarcity of supply, the Government of National Defense took control of the horse market. The implementation of rationing resulted not only in the issue of ration cards but also a general census of all horses.[25]

The authorities increasingly saw horsemeat as a perfect substitute for beef and mutton. Indeed, although its price increased with the siege, horsemeat – fresh or salted – always remained the cheapest meat. From mid-November, its distribution was thus organized along the same principles as those for beef and mutton.[26] This

---

22 *Bulletin de la municipalité de Saint-Denis*, no. 1, 15 November 1870.

23 *Les murailles politiques françaises*, 1874: 246.

24 Decree of 5 November 1870 by the Ministry of Agriculture and Trade.

25 Decree of the Government of National Defense dated 25 November 1870. The requisition was to make it possible to count the horses intended for human consumption. A later requisition decree dated 25 December 1870 transferred the property of horses to the government.

26 Decree of 11 November 1870 by the Ministry of Agriculture and Trade.

also applied to horsemeat offal which was considered as butcher's meat at the end of 1870.[27] The daily ration of horsemeat went from 150g at the end of 1870 to 100g at the beginning of 1871. It is worth pointing out that out of 100g edible meat represented only 80g with the remainder accounted for by bones. Horsemeat was mostly used to prepare a stew that Madeleine Ferrières called 'the main dish of the siege', but sausages made out of horsemeat as well as pork and beef were also eaten.[28]

After the siege, demand for horsemeat increased significantly. Simultaneously, the number of animals killed rose from a few hundred before the siege of Paris to tens of thousands per year. The figure stood at more than 10,000 in 1877 and more than 60,000 in 1910. At the end of the nineteenth century, while beef or pork was often infected with tuberculosis or trichinosis, horsemeat was presented by its defenders as a 'healthy' meat. Doctors insisted on its high content in iron and albuminoidal substances which made it an ideal meat for laborers, but also for patients and convalescents. Hospitals served it in great quantities to their patients.[29] Nevertheless, horsemeat was mostly a 'popular' meat because it was cheap and widely available. There were nearly three million horses in France at the end of the nineteenth century. In Paris, the price for horsemeat was only 60 cents per kilo compared with 1.50 francs for beef or veal in 1882.

Horsemeat as the Beef for the Poor

Benefiting from favorable tax conditions, including exemption from certain taxes, horsemeat butchers multiplied in Paris and in the provinces. According to Leteux '... hippophagy developed primarily in city centres, in areas with a strong labor force' such as Paris, Nord-Pas-de-Calais.[30] At the end of 1908, there were 311 horsemeat butchers and more than 200 stalls in Parisian markets out of a total of 440 butcher's shops in France as a whole. Their geographical distribution was not random. As Ghislaine Bouchet (1993) has demonstrated, they were more numerous in popular neighborhoods. In Paris, the 20th district thus had the greatest

---

27   The maximum price for horsemeat offal sold on 25, 27 and 29 November and 1, 3 and 5 December 1870 in the 10th district specified: '50 grams of kidneys, tongue, liver, heart, brain, will be equivalent to a ration of 50g of ordinary butcher's meat'.

28   Ferrières 2007: 459.

29   The *Larousse gastronomique* 1960 takes up the generally assumed ideas on horsemeat, 'Although horsemeat is less tasty than beef for example, its consumption increases continuously for reasons of economy. Richer in glycogen than other meat, it has a sweet savor which would remind of the roe-deer, in the young and well nourished animals. Since the horse is not likely to catch tuberculosis and not sensitive to the tapeworm's cysticercus, its flesh is recommended in the diets which favor the use of raw meat. It is however necessary to be sure of its quality, because horsemeat preserves badly and it is often difficult to appreciate its freshness. The juice extracted from it is fortifying'.

30   Leteux 2005: 153.

number of horsemeat butcher's shops, 36, followed by the 18th district with 27, the 11th district with 23, the 13th with 22, the 15th with 19, the 17th with 16, and finally the 10th. The 16th district had the smallest number of horsemeat butcher's shops. Only some of the meat was consumed fresh; it was mostly used in delicatessen such as sausages.[31]

In view of the rapid rise in the consumption of horsemeat at the end of the nineteenth century, horse markets close to slaughterhouses were no longer adequate. To meet the increased demand new slaughterhouses were opened. The process began in the 1860s in Pantin (1867) and Villejuif (1866), and the slaughterhouse of Vaugirard, also called Brancion Decroix, opened in 1904.[32] The trade got organized and a butcher's union was established in 1890. It also published also a newspaper, *The Horsemeat's Butcher's Shop of France*. Domestic supplies were no longer sufficient and it soon became necessary to import horses. The first foreign horses arrived in France just before 1914.

A Meat with an Ambiguous Status

Consumed as a substitute for beef, presented as beef or mixed with other meat, everything was done as if horsemeat did not exist by itself, as if there were an absence of identity of what by necessity was called the beef of the poor. One would look in vain for horsemeat recipes at least in the cookbooks written by chefs like Escoffier, Pellaprat and Guégan, or even in housewives' cookbooks.[33] When restrictions returned due to the occupation of France in the Second World War the famous gourmet Edouard de Pomiane gave two horsemeat recipes, the roasted filet and the filet *grand veneur*.[34] Even though it mentioned Arles's dry sausage as a specialty, the *Larousse gastronomique* referred the reader to beef for 'various culinary preparations'. The *Dictionnaire de l'Académie des Gastronomes* (1962) claims that 'Horsemeat, of mild taste and slightly sweet, is far from being worth as much as beef, which it looks like. It only offers more safety: the horse can get neither tuberculosis nor the tapeworm'.

Confusion and fraud went hand in hand for a long time. Indeed, chefs as well as dishonest restaurant owners served horsemeat to their consumers in place of beef or venison. Some supposed pork sausages were in fact made with nothing but

---

31 In France the most well-known sausages are the dry sausage of Arles and of Lorraine. These were also manufactured in other European countries where the consumption of horsemeat started earlier than in France. In Denmark 'spepepolse' were popular.

32 For a long time, this was the largest horsemeat slaughterhouse in France, measuring 40,000 m². Its stables, well ventilated and without access to the bovine stables, could accommodate nearly 300 horses.

33 The great classic *I can Cook* by Ginette Mathiot claims: 'All the culinary preparations being appropriate for beef can be followed for horsemeat'.

34 In his book *Cuisine et Restrictions* (1940: 152–4).

horse or donkey meat. Unscrupulous butchers had been mixing horse or donkey meat with other meats for a long time. This resulted in protests by the trade union of butcher's shops. From 1895 onwards it was possible to identify the presence of horsemeat in a dry sausage thanks to a process developed by two chemists, Edelman and Brautigam.[35] The state intervened on several occasions to fight against fraud and to regulate production and consumption. The Bocher Law of 27 May 1874 defined the breeding conditions and it fixed the number of national studs. There were also ordinances and circulars on slaughterhouses (1873, 1905 and 1920) and the role of veterinarians (1882). While beneficial in their time, these measures were not sufficient to prevent the crisis and contemporary decline.

An Inescapable Decline

After the Second World War, consumption of horsemeat rose slightly, but production quickly became insufficient because of the modernization of agriculture.[36] Motorization had started before the war and it developed rapidly in the 1950s. Tractors replaced horses and the equine population dropped rapidly. Consequently, the supply of animals on the market declined and imports were unable to make up the shortfall. The price of horsemeat increased and gradually caught up with that of beef. The cheap and popular status of horsemeat disappeared.

Indeed, imports actually contributed to the crisis. In 1976 trichinosis, a parasitic disease related to the consumption of raw or undercooked horsemeat was observed for the first time. Since then, eight epidemics caused by the consumption of horsemeat have affected more than 2,000 people and caused several deaths in France. Investigations revealed that in several cases the infection was traced to the consumption of horses imported either from Eastern Europe or the United States. Consumption fell rapidly in the 1980s. It declined by 5 per cent per year throughout the decade and by 9 per cent per year from 1990 to 1995. In 1998, the annual per capita consumption of horsemeat stood at only 0.6 kg, whereas it used to be 26.4 kg. According to Mainsant and de Fontguyon (1986) this decline was due to the absence of horsemeat butcher's shops in the vicinity and the high cost and the taste of the meat. It is also necessary to take into account the motivations of the consumers. Hippophagy was more popular in Paris, in the North and Picardy, with the lower class and young couples with children while the privileged social

---

35   The method consisted in treating a soup of the suspicious product with iodized water. The contact between iodized water and horsemeat soup creates a distinctive purple brown color, which does not occur in other meat soups.

36   In 1954, 90,000 tons of horsemeat were produced and 90,000 tons were consumed. The figures for 1955 were 90,000 tons and 91,000 tons respectively and in 1956, production stood at 85,000 tons but consumption rose to 93,000 tons. In 1956, while France exported 1,078 tons of horsemeat, it imported 8,955 tons and, therefore, there was a deficit of 7,877 tons.

categories refused to consume it. Two thirds of consumers were reluctant to offer horsemeat when they had guests.

Conclusion

Beyond the historical circumstances which clarify the evolution of its consumption, from a rise in the 1870s to its recent fall, the history of horsemeat demonstrates the important part played by representations, beliefs and prejudices in the human diet. The prejudices against horsemeat consumption never completely disappeared as illustrated by the fact that equine butchers never presented anything on their stalls that could remind people of the animal – may it be its skin, its bones or its head. As Leteux put it, horsemeat 'must come forward wearing a mask if it wants to be consumed'.[37] If it was possible at the end of the nineteenth century to ignore secular prohibition, the recent collapse in consumption testifies to persistent resistance, not to mention the failure of propaganda and teaching about nutrition. Even though promotion campaigns continue to underline the nutritional advantages of horsemeat, its high content in iron and glycogen and its low level of lipids, consumers still prefer beef or poultry.[38]

References

Annales d'hygiène publique et de médecine légale [Annals on Public Hygiene and Legal Medicine], June 1885 vol. XIII]

Ghislaine Bouchet. 1993. *Le cheval à Paris de 1850 à 1914.* [Horse in Paris from 1850 to 1914] Genève: Droz, Paris: diff.Champion.

Bizet, L.C. 1847. *Du commerce de la boucherie et de la charcuterie de Paris et des communes qui en dépendent tels que la fonte des suifs, la triperie.* [On Butcher's and Delicatessen Shops in Paris and its Surroundings such as Tripe Shops] Paris: Paul Dupont.

Cinquantenaire des abattoirs de Vaugirard (1904–1954) Historique de l'hippophagie. [Fiftieth Anniversary of Vaugirard's Slaugterhouses (1904–1954) History of Hippophagy] 1954. Paris: Edité par la Fédération nationale et la Chambre syndicale parisienne d'industrie hippophagique, Paris: Imprimerie Genèse.

Decroix, E. 1885. *Recherches expérimentales sur la viande de cheval et sur les viandes insalubres au point de vue de l'alimentation publique.* [Experimental Research on Horsemeat and the Unhealthy Meats in the Public Diet] Paris: Baillière et fils.

---

37   Leteux 2005: 154

38   In 1994–95, a campaign by the horsemeat butcher's union reminded people that horsemeat was good for a balanced diet, strength and energy.

Delavigne, A.E. 1999. Nous on mange de la viande. Approche anthropologique du rapport à la viande au Danemark. [We eat Meat: Anthropological Approach on Attitudes to Meat in Denmark] PhD thesis, Ecole des Hautes Etudes en Sciences Sociales, Paris.

Dictionnaire de l'Académie des Gastronomes. [Dictionnary of the Academy of Gastronomes] 1962. Paris: Aux Editions Prisma, vol. I, 241.

Ferrières, M. 2007. *Nourritures canailles.* [Scoundrel Food] Paris: Le Seuil.

Fierro, A. 1996. *Histoire et dictionnaire de Paris.* [History and Dictionnary of Paris] Paris: Robert Laffont.

Geoffroy Saint-Hilaire, I. 1856a. *Lettres sur les substances alimentaires et particulièrement sur la viande de cheval.* [Letters on Food and Particularly on Horsemeat] Paris: Victor Masson.

Geoffroy Saint-Hilaire, I. 1856b. De l'usage alimentaire de la viande de cheval. [On the use of horsemeat in diets] *Comptes rendus hebdomadaires des séances de l'Académie des sciences*, 43 Mallet Bachelier, Imprimeur Libraire, 455–9.

Geoffroy Saint-Hilaire, I, 1868. *Le Moniteur Universel Journal officiel de l'Empire français*, 200, 18 July, 1074.

Leclerc, comte de Buffon. 1753. *Histoire naturelle, générale et particulière, avec la description du cabinet du Roy.* [Natural, General and Particular History, with a Description of the King's Chambers] Vol. IV. Paris: De Imprimerie royale.

Les Murailles d'Alsace-Lorraine. [The Great Wall of Alsace-Lorraine] 1874. Paris: Le Chevalier éditeur.

Les Murailles politiques françaises. [The Great Wall of French politics] 1874. Paris: Le Chevalier éditeur.

Leteux, S. 2005. L'hippophagie en France La difficile acceptation d'une viande honteuse, [Hippophagy in France: the difficult acceptance of a shameful meat] in Sociologie de l'alimentation: la table, le marché et la nature: *Terrains et Travaux*, revue de sciences sociales, 9.

Mainsant, P. and G. de Fontguyon. 1986. *La demande française de viande de cheval.* [French Demand for Horsemeat] INRA, OFFIVAL.

Mathiot, G. and H. Delage. 1932. *Je sais cuisiner.* [I Know How to Cook] Paris: Albin Michel.

Molinari, G. 1871. L'alimentation de Paris pendant le siège (1871). [Diet in Paris during the Siege] *Revue des Deux Mondes*, 41, 112–23.

Montagné, P. 1960. *Nouveau Larousse gastronomique*, [New Larousse on Food] revised edition by Robert J. Courtine. Paris: Librairie Larousse.

Parent-Duchatelet, A. 1832. Des chantiers d'écarissage de la Ville de Paris, [Quartering sites in the city of Paris] *Annales d'hygiène publique et de médecine légale*, 1(8), Paris: Jean-Baptiste Baillière, 120–21.

de Pomiane, E. 1940. *Cuisine et restrictions.* [Cooking and Restrictions] Paris: Correa.

Wagner, M.A. 2005. *Le cheval dans les croyances germaniques Paganisme, christianisme et traditions.* [The Horse in the Germanic Beliefs, Paganism, Christianity and Traditions] Paris: Honoré Champion.

Chapter 18

The Innovative Power of War: The Army, Food Sciences and the Food Industry in Germany in the Twentieth Century

Ulrike Thoms

Introduction

The German army had gone to war in an enthusiastic mood in 1914; however soldiers returned exhausted and underfed to their hungry families at the end of the war. The experience of the First World War had clearly demonstrated the fragility of Germany's national food economy and, therefore, the National Socialist regime began to plan its food strategy from early on. Very soon after the Nazis came to power they began to subsidize research institutes in the field of agricultural research heavily. A large and comprehensive research programme on breeding, food manufacture and physiology was established. The *Deutsche Gesellschaft für Ernährung* [German Dietetic Association] was founded and funding of the German Research Association was increased. Finally, new journals like the *Zeitschrift für Ernährung* [Journal of Nutrition] and *Zeitschrift für Gemeinschaftsverpflegung* [Journal of Collective Feeding] were established in 1936.[1]

The maintenance and, indeed, increase of performance and health were the leading principles in medicine and research. In this context the soldier's diet was of basic importance for the entire working population and the soldier, further, served as a model for all men. Based on the concept of the survival of the fittest, the ideal soldier was imagined as an athlete – an image exemplified by gigantic sculptures like '*Die Wehrmacht*' [The army] or '*Die Partei*' [The party] by Arno Breker.

It was hoped that physical training and a well-balanced diet during military service would educate young soldiers by practically demonstrating healthy eating habits.[2] Thus research on army food and its composition according to scientific criteria tells us a great deal about how the ideal body and its proper feeding were conceptualized. Nevertheless, army food was not representative; it covered the male population only. Aside from supplying all necessary nutrients (above all proteins and vitamins) it had to fulfil some special demands. Army food should not to be too expensive and preparation should be easy – even in primitive

---

1 Thoms 2006, Thoms 2010.
2 Ziegelmayer 1937a.

conditions. If possible, the food should be precooked and ready to eat, have a small volume and low weight and, finally, provide a high degree of variation. Research on army food was conducted, first, by physiologists, who fixed the biological and physiological needs for a proper and healthy diet and, secondly, by food scientists who introduced some more practical aspects such as food safety. Administrative bodies then were responsible for transport logistics and, finally, for the transformation of foods into tasty dishes.

The Army and the Food Sciences

In fact, theory and practice often collided. Whereas the physiological and chemical research was performed at the *Militärärztliche Akademie* [Military Medical Academy] in Berlin, the daily and practical provisioning was organized by the *Heeresverwaltungsamt* [Army Administrative Office]. The Military Medical Academy was divided into teaching groups one of which conducted physiological research. All the heads and directors of the groups were professors and taught at the Friedrich-Wilhelm-Universität in Berlin, where army physiology had become an obligatory course for all doctors in training.

Despite these efforts it took until 1939 for new regulations on military food rations to be published.[3] The following year saw the foundation of the *Arbeitsgemeinschaft Ernährung der Wehrmacht* [Working Group on Army Food]. In modern terms we may call it an interdisciplinary group, as it comprised officers from all army divisions, scientists and representatives from industry. The group got off to a bad start and lost focus following efforts to tackle the vitamin problem.[4] However, the situation changed when Wilhelm Ziegelmayer (1898-1950) took over the enterprise. Ziegelmayer was a former teacher, who had graduated in chemistry and entered the Army High Command in 1937. Since the 1920s he had been eager to popularize his work. His book *Rohstoff-Fragen der deutschen Volksernährung* [Commodity Questions in the German Diet] followed the official food policy. It argued strongly for German autarchy and pleaded for making use of new foodstuffs. This book was welcomed as an excellent overview in the field and it went through five editions until 1941.[5]

The Working Group on Army Food was occupied with the practical application of scientific findings on army food and dietary composition, including new foodstuffs.[6] Representatives from industry were invited to the meetings, although the military clearly dominated the 19 meetings that took place until 1945. The protocols of the first four meetings, held until 12 December 1941, were published.

---

3   Thoms 2006.

4   See the report on the vitamin-meeting from 1.–3.8.1940 in Munich, Military Archive, Freiburg (hereafter MA), RH 12–23/1642.

5   Ziegelmayer 1936, 1937c, 1939, 1941, Ziegelmayer 1947.

6   Ibid. 3.

They were available in normal bookshops, but further conferences were treated as military secrets and kept confidential. From 1942 onwards the only articles on army food published effectively amounted to propaganda, making it impossible to reconstruct the actual food situation and the deterioration of provisioning.[7] A major topic in the protocols was vitamins, which were regarded as silver bullets.[8] Cases of scurvy were rarely reported, but numerous researchers reported on cases of hypovitaminosis, a latent deficiency which was said to lead to symptoms such as a loss of performance, combined with bad sleep, increased fatigue and a bad mood.[9] The consumption of vitamin C in these cases should help to increase vigour and capacity.[10] These findings resulted in requests for vitamin supplements, which were integrated into the field rations in the form of vitamin drops. The SS went so far as to set up its own production of natural vitamins which were thought to be even more effective.[11]

This practice conflicted with the scientific views of army nutrition physiologists, who were not convinced that the use of vitamin C could increase capacity. Criticizing the 'dangerous and too strong shift in the perspective of recent times' they stressed that the real danger was not vitamin deficiency but under-nourishment and starvation as well as the low protein content of the soldier's diet. These were harsh words, especially in view of the fact that they were published in the first report of the Working Group on the Food Question.[12] Otto Ranke, the group's chairman, spoke frankly of a 'vitamin hype', judging hypovitaminosis to be an unproven state.[13] He prohibited the distribution of vitamin preparations among the staff of the Military Medical Academy, although this was common practice in numerous industrial enterprises in the context of wider '*Vitaminaktionen*' [vitamin campaigns].[14]

To cap this criticism, academy member Konrad Lang (1898–1985) called attention to the fact that the amount of calories provided in British war prisoner camps was higher than in German camps and demonstrated the mistakes in the calculation.[15] Other members of the academy followed him by criticising another wonder weapon – yeast – as an irrelevant contribution to diet. Quantities consumed were simply too small to carry any weight in regard to protein content in the diet. In fact yeast could only play a role as a condiment.[16] These examples may suffice

---

7   See the numerous articles in *Zeitschrift für Heeresverwaltung*.

8   Bächi 2009: 127ff.

9   On the vitamin hype and the vitamin actions of the Deutsche Arbeitsfront, see Thoms 2007.

10   Matthes 1940: 409.

11   Kopke 2005.

12   Piesczek and Ziegelmayer 1942: 218f.

13   See the note in Otto Ranke's war diary, 11 January 1940, MA Freiburg, RH 12–23/1644.

14   Letter 16 March 1944, MA Freiburg, RH 12–23/1640.

15   Letter 10 June 1940, MA Freiburg, RH 12–23/1640.

16   Ranke, 16 June 1941, ibid. RH 12–23/1640.

to demonstrate that the theoretical, scientifically educated physiologists at the Military Medical Army and members of the provision department lived on different planets. The scientists realized very well that the new foods had limited effects on a practical level and that they were used for propaganda purposes. Notwithstanding, they used the situation to promote their professional aims. Without any obvious moral qualms, scientists monitored starving inmates in prisoner of war camps to study the effects of hunger and starvation. Moreover, they conducted feeding trials in order to establish how to restore starving people back to health.[17]

The results of these trials were kept as secret as the real effects of hunger on the German soldiers. The longer the war lasted, the more deficiencies occurred despite all efforts to improve the rations. Military strategists categorically refused to allow any such information to be publicized, simply stating that, 'The topic of hunger is not to occur in the reports'.[18] None of the 600 doctors in the battle of Stalingrad dared to speak of hunger in his reports until June 1942 when army doctors were officially 'puzzled' about the accumulation of oedema cases. At the same time, the problem of hunger and its extent was very well known on the home front as a result of soldiers' letters.[19]

Even in this situation, physiologists continued to do their duty and they calculated the calorie content of the daily rations. They were not squeamish and accepted it as given that soldiers experienced under-nourishment with all the inevitable consequences. Even when the distributed food went below the official rations, they did little more than to recommend reduced physical activity and adequate sleep.[20] At the same time, physiologists took the opportunity to observe the consequences of under-nutrition and to undertake therapeutic trials for the sake of their scientific research.[21] The only time they protested was when the menus were cut down by the administration in June 1944 without having heard physiologists' opinion on the policy.

The situation of industrial food technologists was rather different, because they were much more orientated towards the realities of war. The High Command of the Army and the Army Administration Office closely worked together with industrial enterprises, who received money from the Four-Year-Plan to promote the development of new foods and industrial food research. The army was an attractive client for the industry, because it promised high sales and fair payment. Moreover, supplying the army, which was perceived as a form of quality approval, provided a powerful argument in advertising.

---

17   Thoms 2004.

18   See the protocol of the pre-meeting for the third East-Conference (Osttagung), MA Freiburg, RH 12--23/247.

19   Kehrig 1974: 300, 338, 500, 501--2, Müller 1993.

20   Letter Ranke, Lang, Gemeinhardt to the Teaching Group C, 11 November 1943, MA Freiburg, RH 12–23/1656.

21   Letter Gutzeit to the army's sanitary inspection, 24 March 1943, MA Freiburg, RH 12–23/247.

Although German food technology, which had once been a world leader, had lost ground, the Allies were surprised about the quantity and quality of preserved foods in Germany at the end of the Second World War.[22] There were about 17,000 patents during this time period and 9,540 touched upon food technology. Of these, roughly one third was concerned with food production and two-thirds with processing.[23] The German armed forces proudly announced that they had developed 140 new foods between 1936 and 1940.[24] The leading figures were convinced that progress in this field would not only benefit the army but also the population as a whole.[25] Food technology played an important role in portraying the *Wehrmacht* as a modern army, which relied upon the results of progressive scientific research and engineering. Food journals printed extensive reports on newly invented foods, the new vitamin drops and special rations to placate the home front. A detailed report on Adolf Hitler's inspection of the 'Westwall' was released in 1939. During the visit, Hitler was shown new foods such as crisp bread, dried potatoes and vegetables, field biscuits, whale meat, sausage preserves, army soups from milled rye with milk protein, soybeans and German spices, army chocolate, the new *Scho-ka-Kola*, a chocolate with added caffeine, and fruit drink powder preparations.[26] Some of these foods were cooked in a test meal for the *Führer*, who was also shown the new field cookbook. It is impossible to comment on all the newly invented foods here. The following sections concentrate on three main items, which differed in importance and underwent divergent developments after the war, namely dehydrated foods, soybeans and frozen foods.

Dehydrated Foods

Dehydrated foods had been used in army food and mass catering since the nineteenth century. Instant soups consisting of simple mixtures of seasoned pulse flour, fat and ground vegetables had been integrated into soldiers' iron ration during the Crimean war.[27] A large dehydration industry emerged during the First World War, when the number of newly built factories in Germany reached over 500. As most producers had no experience in this field the products often were of low quality and they acquired the nickname 'hay wire circuit', a verdict that disparaged the product for a long time, despite the fact that a lot of research was

---

22 Pilcher 1947.
23 See Wagner 1949.
24 Neuzeitliche Gesichtspunkte bei der Heeresverpflegung, *Die Deutsche Fettwirtschaft*, 19 (1942): 167–68.
25 See Piesczek 1940, Hellweger 1943.
26 See the list of newly invented foods, Ziegelmayer 1941: 242–4.
27 Teuteberg, Ellerbrock, Spiekermann, Thoms, Zatsch 1990.

conducted to improve quality.[28] Wherever possible, tinned meat and tinned ready cooked meals were favoured for army purposes, especially as these would help to avoid the problems due to inexperienced cooks. A large industry of canned vegetables sprang up after 1900 and consumption rose dramatically. During the 1930s, canned foods were no longer luxury items,[29] but in the face of the Four Year Plan of 1936 and its orientation towards autarchy, the supply of metal became problematic.

Dehydrated foods not only saved precious raw materials, but they had several additional advantages. Firstly, the technology was simple and safe and, secondly, the cost of drying was much lower than that of canning. Storage relied neither on any additional appliances nor energy and, last but not least, only a simple and cheap paper wrapping was required. Moreover, dried vegetables and other foods lost weight and volume during the drying process. For example, 180 g fresh potatoes were reduced to 20 g which saved precious transport capacities. With further processing, it was possible to reduce the volume even more. The only problem was a loss in taste and vitamins during the drying process and in the storage. This was outweighed by the fact that dried foods needed no further preparation; there was no washing, no peeling or cutting and the foods could be used by all. This was a clear advantage for inexperienced cooks, who had to simply stir the food into boiling water.[30] Therefore, dehydration became the army's favoured conservation method and it was strongly promoted from the mid-1930s onwards.[31] The increasing tin shortage assisted this kind of preservation, even if stannous-free cans and so-called black metal cans were developed in order to help overcome the deficiency.[32] The army was proud to dry everything. Potatoes and vegetables were most important, but during the early 1940s dried cheese, meat, milk and even dried vinegar were introduced.[33] Subsequently production figures rose enormously (see Table 18.1).

Soybeans and Vegetarian Burgers

Dried yeast and vegetarian burgers, which were made from a mixture of soybean powder, spices and herbs were meant to improve protein provision. This was not an entirely new phenomenon. Soybeans had been discussed as a cheap source of

---

28   Luithlen 1940.

29   Die Verwendung von Gemüse- und Fleischkonserven in den Armeen der Grossmächte, *Zeitschrift für diätetische und physikalische Therapie* 5 (1902): 612–15.

30   Ziegelmayer 1941: 239f.

31   Ziegelmayer 1936: 8.

32   Entwicklung einer zinnfreien Konservendose, *Der Vierjahresplan* 4 (1940): 279.

33   Ziegelmayer 1947: 470f.

Table 18.1   The Production of Dried Foods in Germany 1939–44 (in tons)

	1939	1945
Dried Vegetables	1,000	25,000
Dried Potatoes	2,500	65,000
Dried Cheese	–	6,000
Dried Meat	–	4,620
Dried Confiture	–	1,500
Dried low-fat Milk	–	4,560
Mixed dried preserves	–	720
Dried Vegetarian Burgers	10	60,000
Dried Yeast	600	18,000–22,000
Crispbread	1,000	30,000

Source: Ziegelmayer 1947: 467.

protein since the end of the nineteenth century.[34] Public interest peaked during the 1930s because soybeans, which made it possible to circumvent the animal stomach in the production of protein, appeared to close the perceived protein gap perfectly.[35] The only disadvantage was that soybeans had to be imported from Japan, China and Manchuria, which charged Germany's foreign currency account. Therefore, a comprehensive breeding and cultivation programme was established. Firms started large advertising campaigns in 1936 and the government promoted the soybean flour under the name '*Edelsoja*' since 1937.[36] Nevertheless, soybeans did not gain any significant importance in private consumption and firms involved registered losses because the soybean kept its image as a surrogate. I.G. Farben recognized soybeans' relevance to army food and it sought to expand their use in mass catering.[37] As a result, the armed forces started cooking experiments, firstly in the navy, where living conditions on submarines made concentrated high energy foods desirable. The small amounts of about 10–15 g per person were a

---

34   See for example Schilling 1914.

35   Ziegelmayer 1937b. Publication numbers clearly peaked between 1930–39, see Shurtleff and Aoyagi, 1991.

36   Schmidt 1944.

37   Drews 2004: 169.

real improvement of the diet, as their protein equivalent to meat was about 3.25 kg per capita and year.

Moreover soybeans and sprouts were used to stretch out meat rations from 1939.[38] Soybean powder was added to industrially processed foods like instant soup, spices, baked goods, chocolate and sausages in order to improve their consistency and to facilitate their production.[39] The well-known firm Maggi used 1,400 tons of soybeans for the production of instant soups. Another firm involved was Wissoll, which among other products, offered a variety of sandwich spreads containing soybean powder. However, the biggest share of soybean consumption was in the form of vegetarian burgers, which had an immense propaganda potential. The public relations department of the armed forces promoted the product, which was shown in an army food show in Leipzig in 1940[40] and subsequently formed part of a permanent exhibition at the Army Administration Office. In 1940, *The Times* reported on the use of soybean products by the German army and commented that the Germans would not have won the campaign against Poland without their help. This was an important strategic victory which was glorified by reports in German papers under titles like '*The Times* is jealous of the wonder bean'.[41]

Frozen Foods

From the technological point of view, deep freezing was nothing new. The technique had been used in the meat trade with Argentina and the fish trade with Norway since the late nineteenth century. In the early twentieth century frozen fish and meat had become cheap sources of animal protein for the working class. Although freezing and transportation tripled the price of the meat, it still was 40 per cent cheaper than domestically produced meat.

With regard to the future development of frozen foods, the USA, which was said to have developed an efficient freezing industry, played an important role. Rudolf Plank (1886–1973), director of the institute for refrigeration technology at the University of Karlsruhe made the American situation known in Germany.[42] The biggest problem was that the frozen food chain was not yet complete. Most German households did

---

38   Borchers 1944.

39   Heeres-Verpflegungsschau auf der Leipziger Messe, *Zeitschrift für die Heeresverwaltung* 5 (1940): 188.

40   Ibid.

41   The original *Times* article was entitled, 'A vital German supply. The magic bean', *The Times*, 23 April 1940. This article was extensively cited in: Die 'Times' und die 'Nazi-Ernährungspille', *Zeitschrift für die Heeresverwaltung* 5 (1940): 164, Die 'Times' ist neidisch auf die Wunderbohne, *Deutsches Ärzteblatt* 70 (1941): 406.

42   Plank 1929. Report no. 2 was published 1938 in Berlin, report no. 3 1950 in Düsseldorf.

not even have a fridge, let alone a freezer. But even though material conditions for mass consumption of frozen food were bad in the 1930s, the future seemed promising and an extensive research program was set up.[43] Most remarkably the German engineers did not develop their own truly German technology at all, as they did in other fields during these times. Instead 1939 American Birdseye-licenses were taken in order to make quick use of technological progress.[44]

The advantages of deep freezing were obvious, especially in the light of vitamin research. There was no loss of vitamins, no laborious and expensive preparation, no change of taste, no need for precious tin cans and no need for preservatives.[45] Therefore deep freezing was massively sponsored under the auspices of the Four-Year-Plan. Although food scientists stressed the importance of vitamin conservation, the frozen food programme did not focus on the conservation of fruit and vegetables. It was above all meat that was frozen and it was argued that dehydration would be a cheap, practical and safe technology for fruit and vegetables. This makes clear that the development of frozen food in fact followed path dependencies and the established routes of the frozen meat trade, on the one hand, and developments in the USA, on the other.[46]

Furthermore fish seemed to be especially promising for the army.[47] Fish was a relatively cheap source of animal protein, which had been promoted for private consumption and in public institutions like prisons or hospitals since the nineteenth century.[48] Germans were slow to embrace fish so that between 30 and 40 per cent of the fishing industry's capacity were unused in the 1930s. Believing that this was due to the bad quality of fish, the Imperial Ministry for Food and Propaganda built up a refrigeration plant in Wesermünde on the Northern coast. In addition, a fishing trawler was constructed, which caught and processed up to 15 tons of fish per day.[49]

Private enterprise had long recognized the possibilities in this new field. As early as 1925, Walter Schlienz had founded the *Kühlfisch Aktiengesellschaft Wesermünde* [Cool Fish Corporation], other enterprises followed such as *Nordsee*, which originally had only been trading fish.[50] The National Socialist state assisted these efforts by subsidizing state institutes as well as private enterprise. Under the personal promotion of Hermann Göring, Andersen Co., a subsidiary of the corporate tobacco group Reemtsma, entered the refrigeration business in 1940

43   See Mosolff 1940.
44   *Gefrier-Taschenbuch* 1940: 14–15.
45   Paech 1938.
46   Plank 1941.
47   Ziegelmayer 1940.
48   Thoms 2005.
49   Ostertag 1933.
50   Hilck and Auf dem Hövel 1979: 8.

and the firm *Solo-Feinfrost* was founded in 1939 under the influence of the High Command of the *Wehrmacht*.[51]

In 1935 the armed forces opened their own Research Institute at Wesermünde under the direction of Rudolf Heiss, a former student of Plank. Plank's Institute became the *Reichsanstalt für Lebensmittelfrischhaltung* [Imperial Institute for Food Preservation] in 1942. These efforts were not only directed towards research but also towards establishing a functioning technology. Despite the high investments, the effort did not become profitable during these years because the richest fishing grounds were closed to the Germans after the outbreak of war. Nevertheless, even during 1943 and 1944 deep freezing was 40 per cent more profitable than conventional marketing. The difficulties of maintaining the cooling chain and of packaging frozen foods were tackled energetically by National Socialist research institutions. Retailers and small shop keepers could not afford expensive freezers, but these were provided on loan by producers.[52]

By contrast, the army had no major problems in this respect. Once it had been decided to freeze only fish and meat, large containers for bulk storage of meat were constructed. These allowed to transport densely packed meat and to keep it for up to 18 days without any further cooling and without any deterioration.[53] Several fish freezing factories were established in Norway in order to provide enough fish for the army and to improve the protein balance in the soldiers' diets. Under the direction of a firm established in France, vegetables were frozen in the Netherlands, in France and Italy to make use of their rich vegetable production.[54]

The comprehensive technological preparations took place during the years 1937–39. In 1940 the freezing industry started with freezing capacities of 200––800 tons per day. Following a slow start, 22,000 tons of goods were frozen in 1940; of these 14,000 were vegetables and fruit and 7,000 fish.[55] As early as 1940, a quarter of the army's meat consumption came from frozen sources.[56] This coincided with research on freezing techniques and packaging. The main aim was to avoid freezer burn and to prolong the possible conservation time. Different varieties of vegetables and fruit were examined with regard to how they would change by freezing in the Imperial Institute for Quality Research of Plants.

---

51   Ibid.: 8f.

52   See letter Reichskommissar for the Unilever-Konzern to State secretary Backe, 1 September 1942, Bundesarchiv Berlin, R 26IV, Nr. 28.

53   Ziegelmayer 1947: 603.

54   Ziegelmayer 1941: 315.

55   Mosolff 1941.

56   Pieczek 1942: 15.

The Collaboration between Scientists, Industry and the Army

The example of freezing has already thrown some light on the collaboration between the army and industrial enterprises. The high command of the armed forces invited industry to build up collaborative research groups. Among others a group which focused on food technology was founded. One of the most well-known members of the group was Dr Oetker. The firm's cooperation went far beyond mere product distribution. Richard Kaselowsky, who had become the owner of this firm by marriage to the widow of August Oetker was a member of the High Command of the Army as well as of Heinrich Himmler's Circle of Friends [*Freundeskreis* Heinrich Himmler].[57] The Oetker syndicate printed not only the German ration cards, but also produced food packaging for the army. Located at Bielefeld the firm had a branch office in Berlin, where plans for the firm's expansion into the East were worked out. Its head was Hans Crampe, who as an expert belonged to the food industry branch of the Four-Year-Plan organization. Moreover, in 1943 the firm founded the Hunsa-Forschungs-GmbH with the Phrix Corporation. Oetker held a third of the shares of this firm which led research and production of synthetic protein from yeast. Other examples of collaborations include Maggi, which was involved in the production of soybean products, Pfanni, which delivered potato products to the army and Nestlé with its new dehydrated coffee Nescafé.

Conclusion

Innovations like these played an important role in Germans' perception of themselves and in the representation of Germany as a modern, technically orientated nation of inventors, researchers and engineers. Therefore, new army foods were used for propaganda purposes throughout the war, both on the home front and directed towards other countries. Considerable effort was directed to this field and, indeed, after the capitulation the Allies were surprised about the high level of the technological development. At the same time, the war delayed implementation of new technologies and wartime shortages contributed to the poor quality of many products.

What happened after the end of war? Undoubtedly, many companies had profited a great deal from the wartime research effort and subsidies of new technology. This is difficult to quantify because business archives do not document these collaborations and their outcome. Wilhelm Ziegelmayer's book published in 1947 provides an attempt to sum up the technological innovations and their possibilities.[58] Looking back, his assessment was realistic with regard to drying and freezing. The importance of the former actually increased during the immediate

---

57   Jungbluth 2004: 133ff, Vogelsang 1972.
58   Ziegelmayer 1947.

postwar years, because the cheapness and simplicity of dried foods were in accord with the disastrous situation at the time. Dried vegetables and especially potatoes continued to play an important role, for example during the Berlin blockade, when they became the most important food. Factories like Pfanni made the best of this situation and developed the production of potato flakes for pancakes and mashed potatoes. Dried vegetables eventually lost importance but they continued to be used in the food industry in a more invisible form, for example as an ingredient in dried soup.

The situation after the end of war was much more negative for the freezing business, which continued to be constrained by persistent limitations of the cooling chain. Moreover, many factories and storehouses had been destroyed. However, in view of the rapid growth in the consumption of frozen foods in the USA, producers remained optimistic. Going back to the well-proven tactic of cooperating with large consumers, they developed the market not so much with products for domestic consumption, but for caterers. In fact, many frozen foods started their postwar revival in canteens and institutional catering. [59] Even today, these account for a large share of frozen food consumption. Other products like soybeans and yeast became almost invisible to the consumer, but they were extensively used as ingredients by the modern food industry, either in the form of emulgators, baking aids or flavourings. In spite of this, today they remain specialized products in the sphere of the exotic and organic food business. Nevertheless, the case of food innovations by the German army demonstrates that changes in provisioning and supply were implemented during the war and the decision to invest in this sector was far-sighted, ultimately producing high profits for all involved.

References

Bächi, B. 2009. *Vitamin C für alle! Pharmazeutische Produktion, Vermarktung und Gesundheitspolitik (1933–1953).* [Vitamin C for All: Pharmaceutical Production, Marketing and Health Policy] Zürich: Chronos.

Borchers. 1944. Untersuchungen über die chemische Zusammensetzung der Sojakeimmasse und Bestimmung des Zusatzes derselben in der Wurst. [Studies about the chemical properties of soy and its suitability in sausage production] *Zeitschrift für Untersuchung der Nahrungs- und Genussmittel*, 87, 154–66.

Drews, J. 2004. *Die 'Nazi-Bohne'. Anbau, Verwendung und Auswirkung der Sojabohne im Deutschen Reich und in Südosteuropa (1933–1945).* [The 'Nazi-Bean': Production, Use and Significance of the Soybean in the German Empire and South Eastern Europe] Münster: LIT-Verlag.

Gefrier-Taschenbuch Herstellung, Bewirtschaftung u. Verbrauch schnell gefrorener Lebensmittel. 1940. [Freezer Handbook: Production, Processing

---

59   Thoms 2005: 284–8.

and Consumption of Frozen Food] Edited by Verein Deutscher Ingenieure. Berlin: VDI-Verlag.

Hellweger. 1943. Wechselwirkung zwischen Wissenschaft und Soldatenkost. [The relationship between science and soldiers' diets] *Zeitschrift für Heeresverwaltung*, 8, 240–42.

Hilck, E. and R. Auf dem Hövel, R. 1979. *Jenseits von minus Null: Die Geschichte der deutschen Tiefkühlwirtschaft*. [On the Other Side of O: The History of the German Freezer Economy] Köln: Deutsches Tiefkühlinstitut.

Jungbluth, R. 2004. *Die Oetkers: Geschäft und Geheimnisse der bekanntesten Wirtschaftsdynastie Deutschlands*. [The Oetkers: Buisness and Secrets of the most famous Economic Dynasty in Germany] Frankfurt a.M.: Bastei-Lübbe.

Kehrig, M. 1974. *Stalingrad. Analyse und Dokumentation einer Schlacht.* [Stalingrad: Analysis and Documentation of a Battle] Stalingrad: Deutsche Verlags-Anstalt.

Kopke, C. 2005. Gladiolen aus Dachau: Das Vitamin-C-Projekt der SS. [Gladiolas from Dachau: the vitamin project of the SS] *Bulletin für Faschismus- und Weltkriegsforschung*, 25/26, 200–219.

Luithlen, H. 1940. Trockengemüse. [Dried vegetables] *Zeitschrift für die Heeresverwaltung*, 5, 47.

Matthes, S. 1940. Sportliche Leistungsfähigkeit und Vitamin C. [Sporting fitness and vitamin C] *Medizinische Welt*, 14, 405–10.

Mosolff, H. 1940. Einsatz der Gefrierwirtschaft für die Vorratswirtschaft. [Use of the freezer economy in provisioning] *Der Vierjahresplan*, 4, 140–43.

Mosolff, H. 1941. Der Aufbau der deutschen Gefrierindustrie. [The development of the German freezer economy] *Der Vierjahresplan*, 5, 396–7.

Müller, R.-D. 1993. Was wir an Hunger ausstehen müssen, könnt Ihr euch gar nicht denken. Eine Armee verhungert, [How we suffered from hunger, you can't imagine: an army is starving] in *Stalingrad. Mythos und Wirklichkeit einer Schlacht* [Stalingrad: Myth and Reality of a Battle] edited by W. Wette and G.R. Ueberschär. Frankfurt a.M.: Fischer-Verlag 131–45.

Ostertag, R. von. 1941. Die Verderbnis von Nahrungsmitteln und ihre Verhütung. [The spoiling of foods and how to prevent it] *Zeitschrift für Fleisch- und Milchhygiene* 44, 41–6.

Paech, K. 1938. Gefrieren von Obst und Gemüse. [Freezing fruit and vegetables] *Vorratspflege und Lebensmittelforschung*, 1, 211–16.

Piesczek, E. 1940. Neuzeitliche Gedanken auf dem Gebiete der Verpflegung. [Modern ideas in provisioning] *Zeitschrift für die Heeresverwaltung*, 5, 29–31.

Piesczek, E. and Ziegelmayer, W., eds 1942. *Ernährung der Wehrmacht, 1. Tagungsbericht der Arbeitsgemeinschaft Ernährung der Wehrmacht*. [Feeding the Army, 1: Daily Report of the Working Group Feeding the Army] Dresden/Leipzig: Steinkopff.

Pilcher, R.W. 1947. German Wartime Food Processing and Packaging. *Food Technology*, 1, 394–403.

Plank, R. 1929. *Amerikanische Kältetechnik*. [American Freezer Technology] Berlin: VDI-Verlag.

Plank, R. 1941. Fortschritte der Kältetechnik. [Advances in Freezer Technology] *Forschungen und Fortschritte* 17, 200–204.

Schilling, 1914. Einiges zur Frage der Ernährung der Gefangenen, speziell zur Verwendung der Soja-Bohne beim gesunden und kranken Menschen. [Aspects on the feeding of prisoners, particularly the soybean with healthy and sick people] *Blätter für Gefängniskunde*, 47, 448–59.

Schmidt, H.W. 1944. Eigelblezithin oder Sojalezithin? [Egg yolk or soy lecithin] *Wochenzeitschrift für Wissenschaft, Technik und Wirtschaft der Bäckerei, Mehl und Brot*, 44, 98–9.

Shurtleff, W. and Aoyagi, A. 1991. *Bibliography of Soya Nutrition, Biochemistry and Medicinal Uses. 5456 References from 2000 B. C. to 1991, Extensively Annotated.* 2 vols. Lafayette, California: Soyfoods Center.

Teuteberg, H. J., Ellerbrock, K.-P., Spiekermann, U., Thoms, U., and Zatsch, A. 1990. *Die Rolle des Fleischextrakts für die Ernährungswissenschaften und den Aufstieg der Suppenindustrie: Kleine Geschichte der Fleischbrühe.* [The Role of Meat Extract in Nutrition Science and the Rise of Processed Soup: A Small History of Meat Stock] Stuttgart: Steiner Verlag.

Thoms, U. 2004. Die 'Hunger-Generation' als Ernährungswissenschaftler 1933–1964 zwischen soziokulturellen Gemeinsamkeiten und der Instrumentalisierung von Erfahrung, [The 'Hunger generation' as nutrition scientists 1933–1964 between socio-cultural similarities and the instrumentalization of experience] in *Verräumlichung, Vergleich, Generationalität: Dimensionen der Wissenschaftsgeschichte* [Space, Comparison and Generalization: Dimensions of Economic History] edited by M. Middell, U. Thoms and F. Uekötter. Leipzig: Akademische Verlagsanstalt, 133–53.

Thoms, U. 2005. Industrialising catering. Technological developments and its effects in the twentieth century, in *Lands, Shops and Kitchens. Technology and the Food Chain in Twentieth Century Europe*, edited by C. Sarasuá, P. Scholliers and L. van Molle. Turnhout: David Brown Book Co., 278–95.

Thoms, U. 2006. 'Ernährung ist so wichtig wie Munition'. Die Verpflegung der deutschen Wehrmacht 1933–45, [Nutrition is as important as ammunition: Feeding the German army] in *Medizin im Zweiten Weltkrieg: Militärmedizinische Praxis und medizinische Wissenschaft im 'Totalen Krieg'*, [Medicine in the Second World War: Military Medical Practice and Research during Total War] edited by W. U. Eckart and A. Neumann. Paderborn: Schöningh, 207–30.

Thoms, U. 2007. Vitaminfragen – kein Vitaminrummel? Die deutsche Vitaminforschung in der ersten Hälfte des 20. Jahrhunderts und ihr Verhältnis zur Öffentlichkeit, [Vitamin questions: German vitamin resarch and its relationship to the public in the first half of the 20th century] in *Wissenschaft und Öffentlichkeit als Ressourcen für einander, Studien zur Wissensgeschichte im 20. Jahrhundert,* [Science and Public as Mutual Resources: Studies in Intellectual History in the 20th Century] edited by S. Nikolow and A. Schirrmacher. Frankfurt a.M: Campus Verlag, 75–96.

Thoms, U. 2010. Ressortforschung und Wissenschaft im 20. Jahrhundert: Das Beispiel der Reichs- und Bundesanstalten im Bereich der Ernährung, [Departmental research and science in the 20th century: The example of imperial and federal agencies in the field of nutrition] in *Jenseits von Humboldt. Wissenschaft im Staat 1850–1990*, [Beyond Humboldt: Science and the State] edited by A. C. Hüntelmann and M. C. Schneider. Frankfurt a.M.: Peter Lang, 27–48.

Vogelsang, R. 1972. *Der Freundeskreis Himmler*. [The Circle of Himmler's Friends] Göttingen: Musterschmidt.

Wagner, K. G. 1949. Die Erfindertätigkeit auf dem Nahrungsmittelgebiet in Deutschland während des zweiten Weltkriegs. [Inventions in the area of food in Germany during the Second World War] *Deutsche Lebensmittel-Rundschau* 45, 194–5.

Ziegelmayer, W. 1939. Der Führer läßt sich die Verpflegung der Festungstruppen vorführen [mit 2 Bildtafeln]. [The Führer is shown army food, with pictures] *Zeitschrift für die Heeresverwaltung*, 4, 212–14.

Ziegelmayer, W. 1937a. Die Wehrmacht als Erzieher zur richtigen Verbrauchslenkung und gesunden Ernährung. [The army as educator on correct consumption and healthy diets] *Zeitschrift für Volksernährung*, 12, 13–15.

Ziegelmayer, W. 1937b. Rohstoff 'Eiweiß': Warum gibt es in Deutschland eine Eiweißfrage? [Raw material protein: Why is there a protein question in Germany?] *Der Vierjahresplan. Zeitschrift für nationalsozialistische Wirtschaftspolitik. Amtliche Mitteilungen des Beauftragten für den Vierjahresplan*, 1, 82–4.

Ziegelmayer, W. 1936, 1937c, 1939b, 1941. *Rohstoff-Fragen der deutschen Volksernährung*. [Commodity Questions in the German Diet] Dresden/Leipzig: Steinkopff [with slightly varying titles].

Ziegelmayer, W. 1940. *Die Bedeutung der Kältetechnik für die Versorgung des Heeres: Beiträge zur Kälte- und Lebensmitteltechnik: Vorträge gehalten auf dem Forbildungskurs Kälte- und Lebensmitteltechnik, veranstaltet vom Verein Deutscher Ingenineure im NSBDT*. [The Significance of Freezing Technology for the Supply of the Army: Contributions to Freezing and Food Technology: Lectures presented at a Course organized by the Society of German Engineers in NSBDT] Berlin: VDI-Verlag, 7–8.

Ziegelmayer, W. 1947. *Die Ernährung des deutschen Volkes: Ein Beitrag zur Erhöhung der deutschen Nahrungsmittelproduktion*. [Feeding the German People: A Contribution to the Increase of German Food Production] 5th totally revised Edition of *Rohstoff-Fragen*. Dresden/Leipzig: Steinkopff.

Ziegelmayer, W. 1940. Die Bedeutung der Kältetechnik für die Versorgung des Heeres, [The significance of freezing technology for the supply of the army] in *Beiträge zur Kälte- und Lebensmitteltechnik. Vorträge gehalten auf dem Fortbildungskursus Kälte- und Lebensmitteltechnik veranstaltet vom Verein deutscher Ingenieure im NSBDT im Januar und Februar 1940*, Berlin: VDI-Verlag, 7–8.

Chapter 19
Conclusion

Rachel Duffett

ICREFH's 11th symposium in Paris 2009 met to discuss two preoccupations of our time: food and war. It is an interest confirmed by the most cursory of glances at the shelves of book stores, the popular press, computer game lists or the television schedules. To date, research on the two topics has developed in separate spheres, with a rich historiography of each but few points where the two have been interlinked.

There is, however, a growing interest in the relationship between food and war, as evidenced by the Imperial War Museum, London's recent exhibition 'The Ministry of Food'.[1] The project was developed to mark the seventieth anniversary of the introduction of food rationing in Great Britain, and was an exploration of the ingenuity and frugality necessitated by the food shortages of the period. As the Introduction to this volume has indicated, the published material available has tended to focus on the civilian aspects of eating in wartime and matters such as the provisioning of armies have received very little attention. The chapters in this volume have sought to bridge the gaps in the existing literature and explore the diversity of the relationships between food and war. It is a wide spectrum, ranging from the starvation and suffering of the 1941–44 siege of Leningrad to the very different impact the Second World War had on Iceland, where the conflict brought new opportunities, both nutritional and commercial.

The pressure war exerts on food supplies can also have a variety of results. Shortages can advance new techniques in food production, such as those relating to frozen food outlined in Chapter 18. They can also stimulate populations to seek new substitutes to fill the gaps left by the absence of traditional staples, as the Slovenians found with the development of ersatz coffee using cumin, dried dandelion roots, grape seed, dried black figs, and acorns described in Chapter 7. Conversely, shortfalls in supplies can cause hungry people to look backwards to older foodways as inspiration to improve a diet impoverished by war. The Czech botanist Karel Domin invoked such eating histories in his 1915 study *Crab Grass: The Forgotten Czech Cereal*. In Chapter 6, Martin Franc describes how Domin promoted the reintroduction of a crop that had all but disappeared from the Czech lands in the decades of relative plenty that had preceded the First World War.

Personal responses to hunger also varied, from the sharing strategies that Alicia Guidonet Riera identifies amongst the Spanish population during the Civil War,

---

1   'The Ministry of Food', special exhibition at the Imperial War Museum London, 12 February 2010 to 3 January 2011.

to the less positive behaviours that Kenneth Mouré describes in his study of the French black market in the Second World War. For many, shortages of foodstuffs precipitated a struggle for survival, and as Mouré indicates 'hunger triggered desperation rather than a spirit of sacrifice'.

Food and War in Twentieth Century Europe illuminates the different experiences across the period; it demonstrates that despite their breadth and diversity there are also similarities that underline the fundamental significance of the subject to societies and cultures regardless of national or chronological divisions.

Priorities

The limitations that war places upon food supplies, whether through the disruption of food imports, the restriction of agricultural production or the diversion of revenue into war materiel, dictates a reassessment of priorities. Shortages meant that not all food demands could be met and some mouths were deemed as being worthier of feeding than others. In all instances, regardless of nation or conflict, it is the army that is given first call on whatever provisions are available, usually at the expense of the civilian diet. In Chapter 3 the swift appropriation of Britain's meat supplies by Brigadier S.S. Long of the Army Service Corps in 1914 was the first indication that the traditional ways of organising military provisioning would be insufficient and a new approach was needed. It was clear that this war would not follow the pattern of earlier conflicts, for example the 1870–71 Franco-Prussian War. The rapid expansion of the armies meant that the feeding of millions of men was not a matter that could be left to chance or individual units; government control of the process was essential.

The feeding challenges that armies faced were huge and those official provisioning records that have survived indicate the vast quantities required. In Chapter 2, Peter Lummel writes of the 1.2 billion kilos of flour, one million cattle and one million pigs transported from Germany to their troops on the Western Front in 1914–1916 alone. Armies developed supply systems from scratch, including the building of huge food warehouses such as the Base Supply Depots employed by the British at French ports, to the construction of the railways essential for the daily movement of rations.

In the First World War, technically innovative systems of distribution such as refrigerated stores and a dedicated train service existed alongside the methods of the past, the horses, mules and wagons. The vestiges of the past were something that the Third Reich was keen to eradicate in its vigorous and scientific approach to feeding its army, as Ulrike Thoms indicates in Chapter 18. The failures of earlier military food policy and the hunger experienced in the later stages of the First World War that Lummel identifies, provided a warning of the consequences of a lack of planning and integration in food policy.

Substitutions

War forces compromises upon consumers, and not least were those made by Jewish soldiers in the German Army during the First World War. As Steven Schouten describes, the dietary rituals of *kashrut* were virtually impossible to follow in the frontline and men were compelled to substitute *treyf* military rations for their preferred kosher foods. This research is evidence of the general inflexibility of army provisioning during this conflict. Alternative diets were not provided. Jewish, and also vegetarian, soldiers in both the British and German armies were expected to eat whatever was available. However, accommodations were made by the British War Office for certain groups, such as the troops serving in the Indian Army who received ghee and lentils as part of their rations.

War tested not only traditional and religious foodways but also the ingenuity of populations and their willingness to move beyond the familiar in the quest for food. Several chapters in the book are testimony to the eating boundaries abandoned in this search. Alain Drouard looks back at the impact of the Franco–Prussian War on the dietary habits of the French, where the 1870–71 Siege of Paris resulted in the widespread consumption of horsemeat. He demonstrates how resistance based upon religious and cultural beliefs was overcome in the face of hunger, and that which had once been regarded as inedible became an established ingredient of the French diet.

Hunger in the Czech lands during the First World War resulted in the exploration of a range of substitutes, a number of which as noted above had their roots in earlier periods of shortage. Here Franc also demonstrates how some of the alternatives suggested were used as a vehicle for anti-German feelings. The response of the Czech press to a Bonn professor's suggestion of serving thistles as a vegetable was as much a reflection of the desire to denigrate the alien behaviours of Germans, as it was a comment upon the palatability of the foodstuff.

Hans Teuteberg indicates how the shortages of potatoes, caused by the failure of the 1916 harvest, had a catastrophic effect on German civilian diet. Shortfalls in the supply of the vegetable were especially deleterious given that it had become an increasingly important constituent of a diet that was short of its traditional mainstay, bread. Turnips, normally used as animal fodder, were introduced in an attempt to fill the gap; the health-related evidence of what became known as the 'Turnip Winter' indicates that the substitution failed to compensate for the lack of calories. Similarly, Adel den Hartog describes the hunger in The Netherlands during the Second World War, where supplies of turnips would no doubt have been welcomed by a population forced to consume flower bulbs in the absence of other foods. The pressure on civilians was such that some risked poisoning in their attempts to secure alternative food supplies, not only through the consumption of bulbs, but also as a result of the willingness to eat unfamiliar fungi or excessive quantities of items such as beechnuts.

Strategies

Several chapters in this volume point to the need to pursue new food strategies not only in widening the types of foodstuffs consumed, but in their sourcing. In her chapter on the hunger experienced during and after the Spanish Civil War, Guidonet Reira explores the ways in which civilians were forced to adapt their behaviour in order to avoid starvation. The oral histories describe not merely the need for resourcefulness in obtaining sufficient food, whether through contacts in agricultural areas or the ability to revert to older systems of exchange founded upon barter, but also the pressure to commit acts that transgressed established codes of morality. The painful memories of looting a railway carriage, or stealing from a hotel endured long after the war had ended, evidence of hunger's power to transform behaviour in ways that are unconceivable in times of stability and relative plenty.

Shifts in accepted morality are also evident in the opportunism of the First World War British soldiers, whose scrounging abilities were legendary. Indeed, the artist John Singer Sargent famously captured the image of two soldiers stealing apples from a French orchard in a painting ironically titled *Thou Shalt Not Steal.*[2] The strategies employed to relieve hunger were not exclusively about theft. In Chapter 4, Schouten illustrates the willingness of Eastern European Jewish families to share their kosher food with German soldiers. Cultural and religious bonds between strangers could be strengthened by war.

In wartime, the main responsibility for enforcing the equitable sharing of limited food lies with governments. A number of the lessons learned in the First World War were brought to bear in the actions undertaken in later conflicts and civilian rationing is a key example. In Britain's case, the government had been slow to introduce rationing in the earlier conflict and a policy of voluntary action was initially adopted. People were exhorted to 'Eat Less Bread', a message conveyed in posters and leaflets and also in sermons that clergyman were requested to preach to their congregations. The strategy failed and in 1918 rationing for a number of essentials had to be introduced. The unpopularity of these measures, mainly due to what was regarded as inequalities in distribution, was important in the development of the system designed for use in the Second World War.

British home front food rationing has subsequently assumed a myth-like status as a paradigm of social justice; a foundation stone of the national unity captured in the notion of the 'people's war'. Yet all was not as popularly imagined. Ina Zweinger-Bargielowska points out in her examination of the experience that shares were not as fair as has been widely believed. The introduction of a flat ration deprived the manual labourer, whose higher energy expenditure was insufficiently compensated for in such a system. In addition, the working classes had less time and space to devote to the growing of additional food supplies. Digging for victory was an option that required resources generally unavailable to those working long shifts in the war effort's factories.

---

2 Held at the Imperial War Museum, London.

Gender as well as class played a significant part in the levels of adult consumption and women, as wives and mothers as well as workers, were called upon to make greater sacrifices, in their role as a buffer between state policy and the domestic sphere. Women's critical role in feeding the civilian population is also explored in den Hartog's chapter, which considers the energies expended in The Netherlands on educating housewives in nutritional science and the possibilities of alternative recipes.

A feature of all wartime economies is the rapid development of a black market. When foods are not available legally there will always be those with the contacts to obtain supplies and those with the means to purchase them. Maja Godina Golija describes how in First World War Slovenia, well-dressed women unashamedly hauled rucksacks of illegally obtained foodstuffs back home from food-finding trips to the countryside. The willingness of a population to participate in black market activity is explained by the figures quoted in Chapter 13 on the Occupied France of 1940–44. Mouré estimates that the quantities of food available in 1944 were approximately half of those on offer in 1938, a shortfall that indicates the necessity of accessing the black market for much of the population.

New developments

This volume indicates that many of the changes to food production and consumption that war brings are often abandoned once peace resumes. A notable exception were the technologies and foods developed for provisioning the soldiers of the Third Reich. Thoms emphasizes the efforts that were devoted to research in this area which resulted in the production of 140 new foods between 1936 and 1940. Some of these had longer term implications for Germany. Improvements in the techniques of producing dehydrated food were to prove essential in sustaining the population in the postwar years and played a critical role in feeding Berlin during the blockade.

War encourages, indeed necessitates, the intervention of governments in their citizens' eating habits, and a number sought to exploit the social and nutritional benefits of communal eating. Teuteberg considers the role of Germany's War Kitchens in the First World War and Peter Atkins explores a similar experience in Britain during the 1940s in his chapter on the development of British Restaurants. From the central policy-making perspective, such institutions had much to recommend them. In particular, the possible economies of scale and the positive impact supervised nutrition could have on the population's health and efficiency. However, as Atkins points out, the British government's enthusiasm was greater than that of its people and the number of British Restaurants opened fell far short of the objective of 10,000 sites. A key problem, whether in 1914-18 Germany or Second World War Britain was that of the social aspects of eating. As Teuteberg states, eating communally had unfortunate associations with the traditional methods used to feed the poor, making it unattractive to many.

The opportunity that changes in wartime diet presented for innovative medical research is an area considered by several authors. Pavel Vasilyev explores the profession's response to the impact of malnutrition on mental health during the siege of Leningrad. Here, war provided doctors with an opportunity to examine an under-researched area of human physiology. The endurance of the starving city contributed to both the Soviet war effort and to an expansion of the body of medical knowledge in this area.

Perhaps the greatest dietary experiment described in this volume is the management of Danish nutrition during the First World War. In Chapter 15, Svend Skafte Overgaard considers 'The Year of Health', as its architect the controversial scientist Mikkel Hindhede termed the period 1917–18. The intervention in the nation's diet appeared to vindicate Hindhede's commitment to a high fibre, low protein diet and the population emerged from the war in good health. Significantly, the changes that Hindhede developed were founded in the austere grain-based diet of his own agricultural heritage, a further example of how the solutions to times of hunger could be found in old habits as well as in new developments.

Denmark's experience was atypical and throughout the book it is clear that it has been the civilian populations that bore the brunt of national food shortages, whether in the blockaded Germany of the First World War or Vichy France in the Second. It is fundamental to the prosecution of war that armies have first call upon food reserves, and Overgaard makes the point that neutral Denmark's success in surviving the exigencies of the First World War was partly due to the fact that 'there were no generals demanding meat for their troops'. Hungry soldiers have a disastrous impact upon morale, both militarily and in its wider context. Thoms notes in her chapter that German army doctors were unwilling to make any reference in their reports to the widespread hunger amongst their troops during the Battle of Stalingrad.

Hunger is a by-product of war, which even soldiers are sometimes forced to endure. A coherent national strategy is critical to the success or failure of feeding civilians in particular; although there are circumstances where whatever the government's understanding and insights, the population would have insufficient food. Occupied nations have little official control over consumption, as Isabelle von Bueltzingsloewen points out in her discussion of the politics of rationing in France during the Second World War. France was powerless in the face of German-enforced hunger. Germany's use of the conquered nation as a source to feed her victorious army and population drained food supplies to levels that had a damaging effect on the health of its people, evident in indicators such as the infant mortality rate. Von Bueltzingsloewen also suggests that control of food was prosecuted as an act of war by Germany. Memories of the widespread starvation of its own population in the First World War lingered. As Teuteberg highlights, Germany relied upon imports for one third of its food supplies prior to 1914 and the impact of the Allied food blockade was ruthless, fuelling its subsequent disregard for the hunger of the conquered.

This volume has amassed a wide variety of research on the significance of food in wartime. It reveals the many similarities in the responses of armies, governments and civilians to sourcing and securing food supplies in difficult and often hungry times. The practicalities of feeding populations from severely depleted resources resulted in the involvement of governments in aspects of nutrition that lay outside their ambit in peacetime. In many nations, hunger was differentiated by class; those with the greatest additional resources, whether money for the black market or time and land for additional vegetable growing, ate better than those without.

Gender was also significant as war brought additional responsibilities to women. The combination of their traditional role as the nurturers of families and their presence as the sole provider in the absence of many men serving in the military resulted in a challenging increase in obstacles to feeding themselves and their dependants. The volume also indicates the advantages of a rural environment, where opportunities for obtaining additional food were considerably greater than those available to urban dwellers.

Wars can neither be waged nor won without adequate food supplies; governments rely upon their citizens, whether on the home or battle front, to display levels of endurance and resourcefulness in the face of deprivation. This book demonstrates that their reliance was well-founded; the diversity of responses is evidence of the willingness of civilians and soldiers to explore every possibility in the efforts to assuage the hunger that inevitably accompanies war.

Index